T5-BCI-230

Entomology
A Guide to Information Sources

ENTOMOLOGY
A Guide to Information Sources

Pamela Gilbert and Chris J. Hamilton

MANSELL PUBLISHING LIMITED

ISBN 0 7201 1680 5

Mansell Publishing Limited, 6 All Saints Street, London N1 9RL

First published 1983

© Pamela Gilbert and C. J. Hamilton 1983

Distributed in the United States and Canada by The H. W. Wilson Company, 950 University Avenue, Bronx, New York 10452

British Library Cataloguing in Publication Data

Gilbert, Pamela
  Entomology.
  1. Entomology—Information services
  2. Entomology—Bibliography
  I.Title   II. Hamilton, C.J.
  595.7   QL 468.2

ISBN 0-7201-1680-5

Typeset by Preface Ltd, Salisbury, Wilts.
Printed in Great Britain.

Contents

Preface

It is not the intention of this work to offer an entrée to the literature of particular groups of insects; to accomplish that purpose would require a volume of perhaps twice this size. Rather the work is intended to be an introduction and source book for entomology by subject instead of systematic group.

Entomological literature, like the insects it covers, is proliferating at an alarming rate. For instance, some 1200–1500 papers are published each year on mosquitoes and mosquito-borne diseases alone. Against this background we can perhaps be forgiven, and possibly even thanked, for selecting the standard texts rather than attempting to be totally comprehensive. We have restricted our subject entries mainly to books, monographs and review articles, but other kinds of publication are included when specially pertinent.

Omission or inclusion of a work does not imply either criticism or recommendation on our part. Generally speaking we have included works that we have found to be of most use, either to ourselves or to our library users. We hope that our selection will prove judicious in practice.

We have tried to make our references as complete as possible, except that we have considered giving the numbers of plates or figures to be unnecessary. Journal titles are given in full, as experience has convinced us that use of abbreviations, even from such a worthy source as the *World list of scientific periodicals* (and supplements), has its pitfalls particularly as regards what a cryptic abbreviation actually stands for. For all entries the information provided is, as far as we can tell, correct at the time of going to press but it is inevitable that some new works or new editions of old ones will be published by the time our book appears. Works that we know to be in press, and that are expected to prove useful contributions in their field, are included.

Acknowledgements

The authors wish to record their grateful thanks to those people who have given so freely of their time and advice in compiling this work. It is impossible to name all those who have been so generous with their help. However, a few must be named for the particular help that was given: Drs R.L. Blackman, D.R. Ragge and P.E.S. Whalley gave most useful advice for particular sections of the book. Mrs M.A. Greiff provided some useful suggestions for the section on common names, and Miss C. Best kindly undertook the arduous task of deciphering and typing large parts of the manuscript.

Dr R.W. Crosskey kindly read much of the manuscript and his comments and advice were most helpful. Miss A. Lum also read part of the manuscript and was particularly helpful with works in oriental languages. Special thanks are also due to Chris Hamilton's wife Veronica, without whose encouragement and forbearance several sections might never have been completed.

Our most particular thanks must go to the staff of our respective libraries who took on extra work on our behalf and who have been so forbearing whilst this work was in progress.

Finally we wish to express our thanks to Mr A.P. Harvey, Head of the Department of Library Services, British Museum (Natural History) and Dr N.C. Pant, Director of the Commonwealth Institute of Entomology, for allowing us to take on the compilation of this volume and for encouraging its completion.

1 Introduction

ENTOMOLOGY PAST AND PRESENT: THE SUBJECT AND ITS LITERATURE

The world of insects is immense. At present there are reckoned to be over 900 000 named species, a staggering figure which means that insects account for three-quarters of the living species in the animal kingdom (and half the species of all living organisms if the plants are considered as well). Notwithstanding this impressive total, entomologists believe that even now – after some 250 years of systematic study – it represents only quite a small fraction of the insect species that actually exist and that should, presumably, be described and named.

There is plenty of evidence that the entomologists are right in their supposition. Every day 'new species' are, so to speak, born in the published entomological literature, and if a few apparent species that have already been described disappear into limbo as 'synonyms' their number is greatly outstripped by that of the soundly based new ones. The requirement for describing new species comes not only from the discovery in underworked faunas or insect groups of morphologically distinct species not previously known, but also from finding that existing 'species' are often complexes of biologically distinct species that look exactly alike. The insects seem especially prone to produce such complexes of sibling species, as they are called, and future descriptions of all these siblings could greatly swell the numbers of species. Routine morphological taxonomy, meanwhile, will continue to contribute its share. One example of recent species description will suffice: in a period of 16 months over 1979–80, whilst a catalogue of flies from the Afrotropical region was in press, descriptions of new species from this region alone appeared at the rate of about 15 per month, or one every two days!

The remarkable adaptiveness of insects has enabled them to inhabit almost all land surfaces, and even to flourish in areas as inhospitable as the polar regions and hot thermal springs. A few are even truly oceanic, spending their whole lives far out to sea. A great many species are of concern to man as agricultural pests, destroying an estimated one-third of the world's potential agricultural yield. Others, like the tsetse-fly and the malarial mosquito, attack man or his animals and transmit disease-causing organisms. Not all insects are harmful, however, and there are many which prey on or parasitize harmful species, and which therefore have an important role in pest management in our increasingly environment-conscious society. Some insects are of even more direct commercial benefit to man, such as the honey bee, the silkworm, the lac insect and the cochineal bug.

The enormous economic importance of insects has served to generate a great wealth of literature on all aspects of their biology and control, and the incredible number of species and their great diversity of form has necessitated an equally vast literature on taxonomic entomology.

One of the great early works on natural history, *Hortus sanitatis* ('Garden of Health'), 1491, contains many references to insects as pests of man, and illustrates – amongst others – the flea, the bed-bug and the locust. Bees, as providers of honey, often featured in early literature, and beautiful examples of butterflies can be seen in many of the early illustrated works, such as church manuscripts and Chinese scroll paintings. The scarab beetle is well known in the hieroglyphic inscriptions of ancient Egypt.

The first work in English to appear about insects was Thomas Moffet's posthumous publication of 1634, *Theatrum insectorum*. The original manuscript (British Library manuscript 4014) has some extremely accurate colour paintings of insects in the margins, although the published edition has only uncoloured wood-cuts. Swammerdam's *History of insects* (1754) has many magnificent engravings of whole insects and of their internal anatomy.

Perhaps the most significant date in entomological publishing is 1758, marked by the appearance of the 10th edition of Linnaeus's *Systema naturae,* from which formal naming of the animal kingdom begins. As the founder of the basic classificatory system and the inventor of the binomial system of Latin scientific names for animals and plants, Linnaeus had great influence on the future of taxonomic names in the whole of entomological literature.

The eighteenth and nineteenth centuries saw an enormous increase in natural history publishing. Major expeditions were exploring all parts of the world, often accompanied by naturalists whose special task was to collect and study the exotic fauna and flora. Many thousands of specimens were brought back by these expeditions. The scientific results started to make their appearance in important multi-volume monographs during the middle and latter part of the nineteenth century. These works contained descriptions of the seemingly endless numbers of new species and were often beautifully illustrated; in fact many have not been surpassed in quality of illustration.

A glance at such volumes, or at any detailed catalogue, shows the way in which the insects tended to dominate parts of these series. For some, like the *Fauna of British India,* the insect parts (insects being so extravagantly numerous) nearly caused their financial downfall. This 'take-over' by the insects was not limited to the nineteenth-century literature; it continues to be true of some of the major faunas today, such as the *Faune de France, Fauna USSR* and *Fauna Hungariae,* which show an almost parallel situation.

The nineteenth century also saw the beginning of the entomological societies. Many of these societies were founded by those whose curiosity about the insects and whose wish to meet their fellow 'Aurelians' (as the early insect hunters were known) were matched by a desire for a prestigious journal to publish their own work. The earliest societies to be formed were the meetings of men of wide interests forming such institutions as the Royal

Society, the Linnean Society and the various learned and philosophical societies which began to arise in various countries throughout the world. The special-interest societies came a little later. Certainly the entomologists were early starters – and if the entomological societies were not publishers at their outset, they soon became so. Although at the beginning few of their members (often called Fellows) could be described as professionals by today's understanding of the term, these 'amateur' contributions to the literature were not inconsiderable. Table 1 is a brief survey of the foundation of the major entomological societies of the world.

The learned societies of Europe were not alone in this growth of publishing. In the late nineteenth century the United States was busy establishing State Agricultural Experiment Stations, all of them publishing reports on their work. Many of these early reports contain important descriptions of insects injurious to crops and livestock and the damage they cause. It was during this period that a new profession, that of the applied entomologist, was born, and that the great names of economic entomology began to appear. Forbes, Riley and Fitch, for example, with their volumes on noxious and destructive pests of areas of the USA, provided early publications that contain many of the original descriptions of these insects – and so are still referred to today. The early appointment of state entomologists in the USA is an indication of the importance of insects to the American economy. Even today, despite advanced pest control, the USA produces an economic entomology literature equalling that of the rest of the world put together.

The burgeoning publications and the plethora of new insect names appearing in the literature together forced the birth of the abstracting and indexing services that we know today. Prior to Linnaeus, Réaumur (1683–1757), the well-known French naturalist, complained of the 'fewness of works in the French language on insects' and that 'even in other languages the number is not that great'. Kirby and Spence are said to have consulted some 262 authors for their popular *Introduction to entomology*, published in 1828. By contrast, the French entomologist Lacordaire (1801–70), himself a prolific writer, said of entomological publications that 'their number is such that they succeed each other so rapidly that one of the principal difficulties for those who cultivate the science [entomology] seriously is to keep track of innumerable publications which flood from all quarters' – a point of view which gains much sympathy today. In 1862 Hagen produced his *Bibliotheca entomologica,* a *magnum opus* listing nearly 5000 authors and some 18 130 titles of works. He listed all known papers on entomology, including those of such early authors as Aldrovandi. The work also carries a major subject index arranged by family and a zoogeographical index. Although this was a unique approach at the time, it set a tradition that has continued. 'Hagen' is still an essential reference work and provides the finest listing of early entomological literature in existence. It is especially valuable because it predates the earliest volume of the *Zoological Record.*

The first volume of the *Zoological Record,* for 1864, carried fewer than 500 references for the Insecta; by 1900 this figure had climbed to 1431. For 1976 some 12 300 references are listed. It is pertinent to point out here that the *Zoological Record* does not list papers on economic entomology. These are covered specifically by the *Review of Applied Entomology, Series A (Agricul-*

Table 1   Foundation of entomological societies throughout the world

1745	First Aurelian Society, London	1872	South London Entomological and Natural History Society
1762	Second Aurelian Society, London	1876	Münchener Entomologischer Verein
1801	Third Aurelian Society, London	1879	Entomologischen Föreningen i Stockholm
1822	Entomological Society of Great Britain	1890	Wiener Entomologischer Gesellschaft
1826	Entomological Club (now largely a dining club – the oldest *active* organization)	1904	Ceskoslovenské Spolecnosti Entomologické
1832	Société Entomologique de France	1906	Entomological Society of America
1833	Entomological Society of London	1907	Société Entomologique d'Egypte
1837	Entomologischer Verein, Stettin	1917	Entomological Society of Japan
1842	Entomological Society of Pennsylvania	1918	Sociedad Entomologica de España
1845	Nederlandsche Entomologische Vereeniging	1922	Sociedade Entomologica do Brasil
1855	Société Entomologique de Belgique	1925	Sociedad Entomológica Argentina
1856	Berliner Entomologischer Verein Deutsche Entomologische Gesellschaft	1937	Entomological Society of South Africa
1859	Societas Entomologia Rossica	1937	Sociedade Brasileira de Entomologia (São Paulo)
1859	Entomological Society of Philadelphia	1938	Entomological Society of India
1862	Entomologischer Verein 'Iris' zu Dresden	1945	Lepidoptera Society of Japan
		1947	Lepidoptera Society of America
1862	Entomological Society of New South Wales	1949	Coleopterists Society
1863	Entomological Society of Ontario	1950	Entomological Society of Canada
1867	American Entomological Society	1951	International Union for the Study of Social Insects
1868	Entomologisk Forening in Danmark	1952	Entomological Society of New Zealand
1868	Società Entomologica Italiana	1957	Japanese Society of Applied Entomology and Zoology

tural) and *Series B (Medical and Veterinary)*, which carried a total of 11 000 abstracts in 1976.

The handling of this increasing amount of published information has become a huge problem. Computerized information retrieval systems have eased the problem to a certain extent, but they do not entirely solve the problem for retrospective searching.

The nineteenth century saw the establishment of the major museums of the world. This is not to say that collections, or 'cabinets' as they were known, were not kept before that date. Many important collections were maintained privately, and their owners nearly always made them available for inspection by any person interested in the subject. The exchange, and more particularly the acquisition, of specimens became almost an obsession with some of these early collectors; even sea-captains were employed and carefully instructed on how to collect specimens during their time ashore in 'foreign parts'. Petiver's collection of insects, for instance, was well known, and indeed many of the specimens were seen and used by Linnaeus for his *Systema naturae* (10th edn); part of his collection still remains intact in the British Museum (Natural History).

Gradually the European countries and the Americans built museums to house their rapidly growing collections. These museums became great national and international centres of excellence, world-renowned for their collections and their publications. As collections grew, staff for their curation and taxonomic research became necessary. Even governments came to realize their importance, as witness the British government's requesting the British Museum (Natural History) to accept collections from Africa for identification. In 1899 Ronald Ross proved that mosquitoes transmit malaria – a turning point in the knowledge of tropical diseases that was to bring forth amongst other things the great explosion of literature in medical entomology. In 1898 the then Secretary of State for the Colonies, Joseph Chamberlain, urged the governors of British-administered territories in Africa to have collections made of 'winged insects in the colony which bite men or animals'. The British Museum (Natural History) in its turn was instructed to examine and classify these insects, and Theobald, specially employed by the Museum, was to write his great monograph on the Culicidae.

The need for the specialist to visit and study collections in countries other than his or her own is well accepted in the scientific community and most institutions make their facilities available to visiting scientists. Many, like the British Museum (Natural History), fulfil a wider role in conjunction with universities and other research bodies by accepting postgraduate students for higher degrees, enabling these students to use the collections and library facilities for their research.

Although entomology may appear to be sharply divided into taxonomic studies on the one hand and economic (including medical and veterinary) studies on the other, the two do to a great extent travel together. Biological research, as often stated, must be built on a foundation of correct identification and classification of the organisms involved. This division in the subject is inevitably reflected in the literature, but many journals carry papers on both aspects.

A large proportion of entomological research is financed by the state at both the national and the international level. Most museums are government-funded. Biological research in the field control programmes in developing countries is frequently financed by overseas aid. Agencies such as the World Bank, the Food and Agriculture Organization and the World Health Organization frequently fund research and offer help through the employment of consultants or secondment of staff from academic institutions or research organizations.

The previously mentioned theoretical division between taxonomic and economic entomology is poorly defined, but can be seen to some extent in the literature. It is perhaps most obvious at the textbook level. Perhaps some 60 per cent of the serial titles have mixed content, carrying papers on both taxonomic or descriptive entomology and economic entomology, whilst others are much more specialized. For example, *Polskie Pismo Entomologiczne* carries papers on pest species and their effects on crops as well as important papers on descriptive taxonomy. However, as Poland supports only the one major entomological journal this is hardly surprising. By contrast, in the USA, where many entomological journals are published, the division is not nearly so 'grey'. The *Journal of Economic Entomology* carries precisely what its title purports, whilst the *Transactions of the Entomological Society of America* publishes descriptive taxonomy. Titles are not always so descriptive of their content. There is a tendency for journals to become more specialized. There has been a gradual rise in the number of journals on Lepidoptera for instance, to the point where the number of Lepidoptera publications of all kinds seems endless. Other examples of specialized journals now being published include *Aquatic Insects, Environmental Entomology* and the *International Journal of Insect Embryology and Morphology*. There is no doubt that the specialized journal aimed at a particular audience is meeting with some success. The great proliferation of newsletters evidences this; although not intended as 'publications' in the strict sense of the term, newsletters are nevertheless important sources of information, serving very specific audiences. They are useful, yet often neglected, sources for names and addresses, fields of research and lists of current literature.

The subscription prices of most journals are now becoming prohibitively high for many individual or private subscribers. More and more the researcher comes to rely on libraries, either national, university or institutional, and these libraries in turn are often penalized by having to pay ever higher prices. Few libraries of the world can be relied on to carry exhaustive collections on all aspects of a subject, and some are inevitably much more complete than others.

Large national libraries rarely lend their material, and it is usual for the learned societies to lend material only to their own members or Fellows. However, it is unlikely that anyone will be denied the possibility of using most of these libraries for study purposes, and many countries have an efficient national lending scheme that can if necessary be extended internationally.

Libraries such as those of the British Museum (Natural History), the Muséum National d'Histoire Naturelle in Paris and the American Museum of Natural History, which have collected subject literature over a long period, can usually provide bibliographic fulfilment. This is particularly true for the older literature. Unlike researchers in either the physical or, to some extent, medical sciences, those working in the biological and zoological sciences (including entomology) need to make constant reference to the older literature. Early descriptions of insects are essential to further research, along with the need to trace and examine collections of type-material. It is imperative for taxonomy, because of its inextricable link with zoological nomenclature, to carry its past into its present more than any other branch of entomology (or indeed of science generally); and not simply as a historical curiosity as in some other subject areas.

In a high proportion of libraries (and this includes libraries in institutions, university departments, etc.), entomology is treated as a small subsection of zoology. Exceptions to this are certain large research departments such as those found in the British Museum (Natural History) or the Smithsonian Institution. Relatively few centres anywhere in the world maintain large separate collections of books devoted entirely to entomology. However, most libraries can usually offer the reader some of the more standard works or perhaps general guides. Blackwelder's *Taxonomy* (1967) is a good example of a more general text invaluable to the entomologist. In some 40 pages of the early part of the volume he offers the reader an excellent and succinct guide to the various kinds of literature available to the researcher, and many of the examples given in the text are relevant or indeed refer specifically to the field of entomology.

Winchell's *Guide to reference books* (1951) lists many basic reference works. Its author–title–subject index gives quick reference to many broad subject headings such as scientific societies, dictionaries, etc. However, it must be said that it is not strong on entomology. For this subject, a better general source is *Walford's Guide to reference material*. Volume 1 has appeared in a new (4th) edition (1980). Many of the references have been updated and the entomology chapter gives some useful basic reference sources.

Several guides to the literature of the biological sciences have been published. In general these are student manuals aimed at teaching the student how to use his or her library. One of the best of these is the *Guide to the literature of the life sciences* by Smith and Reid (1972). This is an excellent successor to Smith and Painter's *Guide to the literature of the zoological sciences* (7th edn, 1967). Aimed primarily at the college student with library exercises included, it gives a good entrée to the literature even if again entomology seems to take a back seat!

A good list of titles will be found in Kerker and Murphey's *Biological and biomedical resource literature* (1968). It is mostly a title listing, but the first two chapters are descriptive.

Although it lists mainly by author and title and has very little descriptive material, Rohlfien's paper in *Beiträge zur Entomologie* (1977) is an excellent bibliography of entomological bibliographies covering the years 1920 to 1970. Although many of the titles of papers listed are self-descriptive, it is a pity that the book does not give more detail on the content or coverage of some of the papers.

A work that should not be ignored is Arnett's *Entomological information, storage and retrieval*. Published in 1970 as a teaching manual, it is a mine of useful information; it was not intended as an exhaustive study of the subject, but is nevertheless helpful, particularly some of the appendices (which include lists of entomological suppliers of equipment, dealers and specialist booksellers).

Chamberlin's *Entomological literature and nomenclature* (1952) has for many years been one of the few books specifically on the entomological literature; parts of it are still of great use, but others are now outdated.

Trauger, Shenefelt and Foote's 1974 paper in the *Bulletin of the Entomological Society of America* is an excellent if brief guide to the searching of entomological literature. There is an amazing amount of useful information packed into this relatively short article.

Bottle and Wyatt's 1966 book, *The use of biological literature,* has become well known and is much quoted in information work. The chapter on zoology, however, by D.M. Archer, almost dismisses entomology from the biological sciences: 'Entomology is now a science in its own right'. A short paragraph indicates the 'vastness of the literature' and suggests three volumes as useful works.

The 'art of the newsletter' is well exploited in the field of entomology, and we have devoted some space to this subject area. Sadly, newsletters are often overlooked, although they are one of the most fruitful sources of current information on research. They often give names and addresses of workers in a narrow subject field, together with details of their particular research. While many of the abstracting and indexing services are lagging two or more years behind the published literature (which is itself often less than topical), the newsletter can be most helpful in tracing the current literature. Biographical notices are also well represented. We have always found newsletters invaluable, and their producers are seldom as well appreciated by their readers as perhaps they should be.

Perhaps it might be appropriate here to enter a plea. So many subscribers and specialists tend to treat their newsletters as ephemera quickly to be discarded (or certainly buried on the work bench). The British Museum (Natural History) is trying to make a complete holding collection of these newsletters – which are not always easy to obtain or even to find out about. As new ones are started it would be helpful if their editors could inform major libraries. We (particularly P.G.) would like to know of them.

Entomology is well served with works to aid the researcher with his or her bibliographic problems. Such works are frequently the result of much painstaking work and research and often provide shortcuts to many of the researchers' information needs. Regrettably, however, many of those who would benefit from such works are unaware of their existence and it is our hope that this volume will remedy this situation.

We conclude this brief introduction with the comforting thought that the entomological librarian is unlikely to go out of business in the near future. To judge from recent performance, there is no likelihood that entomologists will lose their propensity for writing up their observations, and seeking outlets for their publication. A few journals will wither and die, no doubt, but others will spring up to cope with the publication demands of an expanding and changing science.

It is the responsibility of librarians and information officers to cope with the ever more voracious appetite of entomologists for information, and this volume is an attempt to point them in the right direction. With it, the entomologist can, we hope, indulge in a little 'do-it-yourself' information retrieval.

References and further reading

Arnett, R.H. 1970. *Entomological information storage and retrieval.* 210pp. Bio-Rand Foundation; Baltimore.

Blackwelder, R. E. 1967. *Taxonomy: a text and reference book.* 698pp. John Wiley; New York.

Blanchard, J.R. and **Otsvold, H.** 1958. *Literature of agricultural research.* 231pp. University of California Press; Berkeley and Los Angeles.

Bonfanti, C. (Ed.). 1979. La investigación bibliografica y la comunicación tecnica. *Alcance Revista de la Facultad de Venezuela* **28**: 1–281.

Bottle, R.T. and **Wyatt, H.V.** 1966. *The use of biological literature.* 286pp. Butterworths; London.

Bottrell, D.G., Huggaker, C.B. and **Smith, R.F.** 1976. *Information systems for alternative methods of pest control.* 42pp. UC/AID Pest Management and related environmental protection project, University of California; Berkeley, California.

Bridson, G. and **Harvey, A.P.** 1971. A checklist of the natural history bibliographies and bibliographical scholarship 1966–1970. *Journal of the Society for the Bibliography of Natural History* **5**: 428–65.

Bridson, G. and **Harvey, A.P.** 1973. A checklist of the natural history bibliographies and bibliographical scholarship 1970–1971. *Journal of the Society for the Bibliography of Natural History* **6**: 263–92.

British Museum (Natural History). 1981. *Report of the British Museum (Natural History) 1978–1980.* 167pp. British Museum (Natural History); London.

Chamberlin, W.J. 1952. *Entomological nomenclature and literature.* vii + 141pp. W.C. Brown, Dubuque; Iowa.

Davis, E.D. 1981. *Using the biological literature: a practical guide.* 272pp. Marcel Dekker; New York.

Dupuis, C. 1954. Les sources bibliographiques de l'entomologiste – Pte 1. *Cahiers Naturalistes. Bulletin* **10** (Suppl.): 77–112.

Edwards, P.I. 1971. The general pattern of biological information. *Biological Journal of the Linnean Society* **3**: 169–72.

Edwards, P.I. 1971. Information and data centres in the biological sciences. *Biological Journal of the Linnean Society* **3**: 249–51.

Edwards, P.I. 1971. List of abstracting and indexing services in pure and applied biology. *Biological Journal of the Linnean Society* **3**: 277–86.

Edwards, P.I. 1971. List of libraries in the field of pure and applied biology. *Biological Journal of the Linnean Society* **3**: 173–88.

Foote, R.H. 1978. Museums, museum data, the real world – the taxonomic connection. In Romberger, J.A. (Ed.). *Biosystematics in agriculture*, vol. 2 pp. 261–74. John Wiley; New York.

Foote, R.H. and **Zidar, J.** A preliminary annotated bibliography of information handling activities in biology. *Journal of the Washington Academy of Sciences* **65**: 19–32.

Gressitt, J.L. 1980. Thoughts on entomological editing and publishing. *Entomologia Generalis* **6**: 379–81.

Hammack, G.M. 1970. *The serial literature of entomology: a descriptive study.* 85pp. Entomological Society of America; Beltsville, Maryland.

Hills, S.J. 1972. A review of literature on primary communication in science and technology. *Aslib Occasional Publications* No. **9**.

Kerker, A.E. and **Murphey, H.T.** 1968. *Biological and biomedical resource literature.* 226pp. Purdue University; West Lafayette, Indiana.

Lilley, G.P. (Ed.). 1981. *Information sources in agriculture and food sciences.* 603pp. Butterworths; London.

Rohlfien, K. 1977. Bibliographie entomologischer Bibliographien (1920–1970). *Beiträge zur Entomologie* **27**: 313–79.

Sabrosky, C.W. 1956. Entomological societies. *Bulletin of the Entomological Society of America* **2**: 1–22.

Sawyer, F.C. 1955. Books of reference in zoology, chiefly bibliographical. *Journal of the Society for the Bibliography of Natural History* **3**: 72–91.

Smith, R.C. and **Reid, W.M.** 1972. *Guide to the literature of the life sciences.* 166pp. Burgess; Minneapolis, Minnesota.

Trauger, S.C., Shenefelt, R.D. and **Foote, R.H.** 1974. Searching entomological literature. *Bulletin of the Entomological Society of America* **20**: 303–15.

Troyer, D.L., Kellogg, M.G. and **Andersen, H.O.** 1972. *Source book for biological sciences.* 176pp. Collier Macmillan; New York.

Walford, A.J. (Ed.). 1980. *Walford's Guide to reference material,* vol. 1. 4th edn. 697pp (pp. 243–247, Entomology). Library Association; London.

Winchell, C.M. 1951–. *Guide to reference books* (and supplements). American Library Association; New York.

Whitehead, P.J.P. 1971. Storage and retrieval of information in systematic zoology. *Biological Journal of the Linnean Society* **3**: 211–20.

Wyatt, H.V. 1967. Research newsletters in the biological sciences – a neglected literature service. *Journal of Documentation* **23**: 321–27.

HISTORY OF ENTOMOLOGY

It is useful and indeed sometimes necessary to have some knowledge of the history of a subject. Entomology is certainly no exception; it has been part of man's past, insects always having been involved in his agricultural economy and his medical health. The subject is well served by published works on its history both in the general and the particular. For the most part these works are well indexed and with good bibliographies capable of leading the user to further reading.

1 Bodenheimer, F.S. 1928. *Materialien zur Geschichte der Entomologie bis Linné.* 2 volumes. Dr W. Junk; Berlin.

An excellent work on the history of entomology, often quoted as the 'standard' history. Each chapter has a short bibliography, but on the whole the referencing is poor. A good index to both the volumes is to be found at the end of vol. 2.

2 Busvine, J.R. 1976. *Insects, hygiene and history.* 262pp. Athlone, London.

Although much of the work is devoted to medical entomology there is a strong bias towards the history of the subject. The work has a most useful bibliography, which includes references to many journals and books outside the field of entomology.

3 Chou, I. 1980. *A history of Chinese entomology.* 213pp. Entomotaxonomia; Shaanzi, China.

The work is written in Chinese, but there is a substantial English summary, which includes all the main events of any importance.

4 Cushing, E.C. *A history of entomology in World War II.* 117pp. Smithsonian Institution; Washington, DC.

The work briefly indicates the importance of research work carried out during the war years aimed at eliminating pests of man.

5 Essig, E.O. 1931. *A history of entomology.* x + 1029pp. Macmillan; New York.

The volume is mostly concerned with American entomology although some internationally known figures are covered. It has a good biography section with many photographs. Chapter 10 is a chronological table showing the development of entomology in relation to the history of other sciences.

6 Goidanich, A. 1975. Uomini, storie e insetti italiani nella scienza del Passato. *Redia* **57**: 1–509; **58**: 511–1060.

A good descriptive work on Italian entomology. It is well illustrated and has much biographical information, especially on very early entomologists and collectors.

7 Gressitt, J.L. and **Szent-Ivany, J.J.H.** 1968. Bibliography of New Guinea entomology. *Pacific Insects Monograph* **18**: 1–674.

A very full bibliography (6140 references) of the scattered literature on New Guinea and nearby islands, including the Bismarck Archipelago and the Solomon Islands. Entries are arranged alphabetically by author and are indexed by subject and by major insect groups. Covers the literature up to 1966, with a very few references from 1967.

8 Henriksen, K. 1921. Oversigt ove Dansk entomologis historie. *Entomologiske Meddelelser* **15**: 1–288.

The emphasis is on biographical material. Contains many photographs of well-known Danish entomologists.

9 Howard, L.O. 1930. A history of applied entomology. *Smithsonian Miscellaneous Collections* No. **84**: 1–564.

International in scope, but with some bias towards American economic entomology. The subject has advanced considerably since the book's time of publication, but for early work this is an excellent account.

10 International Bee Research Association. 1979. *British bee books. A bibliography 1500–1976.* 266pp. International Bee Research Association; Gerrards Cross, Buckinghamshire, England.

Among the first books to be published in England were books on beekeeping. This is a most useful bibliography and source book for the literature of this subject. The main body of the work is a chronological bibliography 1500–99;

1600–99; 1700–99; 1800–49; 1850–99; 1900–26. 1927–76 is an author bibliography. The work also includes manuscripts before 1500. Library holdings are indicated among the annotations.

11   Neave, S.A. 1933. *The history of the Entomological Society of London, 1833–1933.* 224pp. Royal Entomological Society of London; London.

Published to mark the Society's centenary in 1933, the volume traces its history and development, and contains some biographical information on older Fellows of the Society. A list of the Fellows elected since its formation is included.

12   Nordenskiöld, E. 1949. *The history of biology: a survey.* Translated from Swedish by L.B. Eyre. 629pp. Tudor, New York.

Although a general history of biology, it contains information on the history of entomology, with some excellent biographical sketches of individuals.

13   Ordish, G. 1974. *John Curtis and the pioneering of pest control.* vii + 121pp. Osprey; Reading.

This small volume gives some important facts on the history of insect pest control in its early days. It is a good concise account of its early history, not found elsewhere.

14   Osborn, H. 1937. *Fragments of entomological history: including some personal recollections of men and events.* 393pp. Published by the author; Columbus, Ohio.

The emphasis is the history of American entomology. Pages 337–80 contain photographs of entomologists.

15   Osborn, H. 1946. *Fragments of entomological history,* Part II. 232pp. Privately printed; Columbus, Ohio.

This is a supplement to the above volume. Pages 191–226 contain photographs of entomologists.

16   Reigert, P.W. 1980. *From arsenic to DDT: a history of entomology in western Canada.* xii + 357pp. University of Toronto Press; Toronto.

A first attempt to collect in one place the history of entomology in any part of Canada. The volume is in the main a history of applied entomology and concentrates on the insects; there is very little information on entomologists.

17   Smith, R.F. Mittler, T.E. and **Smith, C.N.** (Eds). 1973. *History of entomology.* Annual Reviews; Palo Alto, California.

Twenty chapters on the history of entomology, international in scope and of a most comprehensive nature. Each chapter is complete in itself, taking on the format of the review article, and has a very full bibliography. The whole volume has a good subject index. It is particularly good for early entomology in Eastern Asia, details of which are difficult to find elsewhere.

18   Weidner, H. 1967. Geschichte der Entomologie in Hamburg. *Abhandlungen und Verhandlungen des naturwissenschaftlichen Vereins in Hamburg* **9** (Supplement): 1–387.

A mass of useful information of the history of European as well as simply German entomology.

19 Weiss, H.B. 1936. *The pioneer century of American entomology.* 320pp. Privately published by the author.

Although this book is mostly about American entomology, the author relates this to work being done outside America during the period covered, and uses three chapters of the work to look at pre-nineteenth century entomology. The work is really no more than a typescript, and had a very restricted circulation; it is worth trying to find a copy, as it contains much useful information on early American entomology.

20 Yepez, F.F. 1978. Contribución a la historia de la entomología en Venezuela. *Alance Revista de la Facultad de Agronomía, Universidad Central de Venezuela* **26**: 11–27.

A brief history, with some accent on economic entomology. Contains a good bibliography.

EARLY LITERATURE

Early literature is often the most difficult to trace or check; however, entomology is well served for reliable and well-indexed catalogues or bibliographies of its early literature.

21 British Museum (Natural History). 1903–15. *Catalogue of the books, manuscripts, maps and drawings in the British Museum (Natural History).* 5 volumes. British Museum (Natural History); London.

22 British Museum (Natural History). 1922–40. Supplement to the above 5 volumes. 3 volumes (6–8). British Museum (Natural History); London.

A particularly valuable catalogue for holdings of old and rare works on natural history. It is reliable for the listing of parts and dates. Many works issued in parts are itemized and dated and many of the entries are cross-referenced.

23 British Museum (Natural History). 1980. *Serial publications in the British Museum (Natural History) library.* 3 volumes. British Museum (Natural History); London.

A good source of information for natural history serial titles as well as entomological titles. Essential for a lot of the early literature. The holdings of this library are unique in the field of natural history (including entomology), making this list an excellent finding source. This is the last edition to be published in book form; future editions will be issued in microfiche.

24 Hagen, H.A. 1864. *Bibliotheca entomologica.* 2 volumes. W. Engelmann; Leipzig.

An alphabetical list by author of papers up to 1863. It also contains a good

subject index. The work is out of print, but a reprint has been published by
Wheldon & Wesley Ltd (entry 299).

25 Horn, W. and **Schenkling, S.** 1928–9. *Index litteraturae entomologicae;*
Ser. 1, *Die Welt-Literatur über die gesamte Entomologie bis inklusive
1863.* 4 parts. W. Horn; Berlin.

Virtually a reissue of Hagen, but it does have many additions.

26 Gaedike, R. and **Smetana, O.** 1978. Ergänzungen und Berichtigungen
zu Waltern Horn und Sigmund Schenkling: 'Index litteraturae
entomologicae'; Ser. 1, 'Die Welt-Literatur über die gesamte Entomologie
bis inklusive 1863'. Tl I: A–K. *Beiträge zur Entomologie* **28**: 329–436.

First supplement to the work by Horn and Schenkling.

27 Derksen W. and **Schieding, U.** 1963–75. *Index litteraturae
entomologicae*; Ser. 2, *Die Welt-Literatur über die gesamte Entomologie
von 1864 bis 1900.* 5 volumes. Akademie der Landwirtschaftswissenschaften der Deutschen Demokratischen Republik; Berlin.

These volumes index the literature, continuing from Horn and Schenkling
(1864) to 1900. Volume 5 gives a complete list of abbreviated titles used in
the work with their full journal titles. There is a good subject index by insect
family giving author references and a zoogeographical index.

28 Royal Entomological Society of London. 1980. *Catalogue of the library
of the Royal Entomological Society of London.* 5 volumes. G.K. Hall;
Boston, Massachusetts.

A straight photoprint of the card catalogue of the Society. The holdings of this
library are very rich in rare and privately printed material; these are often
difficult to trace, thus making this catalogue extremely useful. The library is
open for consultation to Fellows of the Society only, but it is also a back-up
library for the British Library Lending Division.

29 Zoological Society of London. 1864– . *Zoological Record, Insecta.*

Published in parts, the *Insecta* part of this indexing service is a record of the
literature published annually since 1864. It carries an author and subject and
systematic index to the literature. Full details of the publishing record of the
Zoological Record are to be found in the section 'Secondary journals' (entry
946).

INSECTS IN ART, LITERATURE AND GASTRONOMY

30 Bastin, H. 1954. *Freaks and marvels of insect life.* 248pp. Hutchinson;
London.

A semi-popular book on the 'strange habits of insect life'. The work describes
some of the curiosities that are often quoted with little explanation – here the
author attempts to provide such explanation.

31 Bodenheimer, F.S. 1951. *Insects as human food.* 352pp. Dr W. Junk; The Hague.

The work brings together a lot of scattered information on the use of insects for nutritional purposes. A good bibliography is provided, including many early travel books, where a lot of these early observations were first published. 'Bodenheimer' has become the basic textbook on this subject.

32 Clausen, L.W. 1954. *Insect fact and folk-lore.* 194pp. Macmillan; London.

A semi-popular book on insects that tells something of the myths and their origins that have grown up around the study of entomology.

33 Holt, V.M. 1885 (reprinted 1967). *Why not eat insects?* 99pp. E.W. Classey; Faringdon, Oxfordshire, England.

A curiosity – but the first work to be written on the subject. It gives a little of the history of insects as food.

34 Kádár, Z. 1978. *Survivals of Greek zoological illuminations in Byzantine manuscripts.* 138pp. 232 plates (10 coloured). Akadémiai Kiadó; Budapest.

Many of the plates include insects.

35 Keller, O. 1913. *Die antike Tierwelt.* 2 volumes. W. Engelmann; Leipzig.

Volume 2 contains a good review of insects and their place in art etc. The work contains no bibliography.

36 Klingender, F. 1971. *Animals in art and thought to the end of the Middle Ages.* 580pp. Routledge & Kegan Paul; London.

A good survey of this early material. Many insects are included in the illustrations, particularly early illuminated manuscripts. A very good bibliography is included.

37 Lenko, K. and **Papavero, N.** 1979. *Insetos no Folclore.* 518pp. Conselho Estadual de Artes e Ciências Humanas; São Paulo.

South America is very rich in insect folklore, much of which is described in this volume (although the book is not exclusively on South America).

38 *Oxford dictionary of quotations.* 1979. 3rd edn. 907pp. Oxford University Press; Oxford.

The larger portion of this work is an index by keywords. By using this keyword index – mostly vernacular names – it is possible to find references to quotations on insects in the literature. The reference shows the author's name (usually abbreviated), followed by page, item and number.

39 Pinney, R. 1964. *The animals in the Bible: the identity and natural history of all the animals mentioned in the Bible.* 227pp. Chilton Books; Philadelphia.

A number of insects are included and some photographs.

40 Schimitschek, E. 1977. Insekten in der bildenden Kunst in Wandel der Zeiten in psychogenetischer Sicht. *Veröffentlichungen aus dem Naturhistorischen Museum, Wien* **14**: 1–119.

Well-illustrated work showing many examples of the use of insects in art, both early and modern. There is also a good bibliography.

41 Smit, F.G.A.M. 1978. *Insects on stamps.* 74pp. Published by the author; Tring, Hertfordshire, England.

A cross-referenced checklist under country of issue.

42 Strom, H. and **Lewy, L.H.** 1968. *Animals on stamps.* 383pp. Philart Productions; London.

Originally published in German. Many insects are included, some illustrated. They are listed geographically by country; chronologically by date of issue and then by animal group within each year.

43 Taylor, R.L. 1975. *Butterflies in my stomach or: insects in human nutrition.* 224pp. Woodbridge Press; Santa Barbara, California.

An interesting book, written by a scientist but intended for the general reader rather than the specialist. Includes much useful information on the actual and potential use of insects as food, and tables showing the nutritional values of various insects compared with those of more conventional foods. Also includes a fairly extensive bibliography.

2 Naming and identification of insects

TAXONOMY AND NOMENCLATURE

Much of the research work and publication in entomology embody, implicitly or explicitly, the findings and opinions of taxonomists on the identities and names of the insects concerned. Classification and naming therefore have an underpinning function in entomology far beyond the immediate interests of the taxonomists themselves. This section cites selected general works on taxonomy and its principles, and works on zoological nomenclature. Nomenclature is a complex subject with its own vocabulary and strict rules to be followed. Table 2 lists insect orders and their common names.

For works on taxonomic identification, faunas, etc., *see* the section 'Identification and faunistics' (p. 24).

44   Advisory Board for the Research Councils. 1979. *Taxonomy in Britain. Report of the review group on taxonomy set up by the Advisory Board for the Research Councils under the chairmanship of Sir Eric Smith, FRS.* 126pp. HMSO; London.

Excellent background information on the taxonomic work being done in the UK, with some recommendations for its future develoments.

45   Blackwelder, R. E. 1967. *Taxonomy, a text and reference book.* 668pp. John Wiley; New York.

An important book which should be familiar to all students. It contains a good guide to the literature and is especially valuable for its detailed synthesis of many of the complex issues in nomenclature.

46   Commission on Biological Nomenclature. *Enzyme nomenclature.* 433pp. Elsevier; Amsterdam.

47   Committee of European Science Research Councils. 1977. *Taxonomy in Europe.* 95pp. Strasbourg. (European Science Research Council Review No. 13.)

Summarizes taxonomic work in Europe, both required and in progress. It lists the major taxonomic resources (i.e. museums, research institutes) in Europe. The work is intended as an interim report.

Table 2 Alphabetical list of insect orders in current usage

Coleoptera	Beetles
Collembola	Springtails
Dermaptera	Earwigs
Dictyoptera	Cockroaches, mantids
Diplura	Diplurans
Diptera	Flies (two-winged)
Embioptera	Web spinners
Ephemeroptera	Mayflies
Grylloblattodea	Rock crawlers
Hemiptera	Plant bugs, etc.
Hymenoptera	Ants, bees, wasps, sawflies, etc.
Isoptera	Termites, white ants
Lepidoptera	Butterflies and moths
Mallophaga	Biting lice, bird lice
Mecoptera	Scorpion flies
Neuroptera	Lacewings, alder flies, snake flies, antlions, etc.
Odonata	Dragonflies
Orthoptera	Grasshoppers, locusts, crickets
Phasmida	Stick and leaf insects
Plecoptera	Stoneflies
Protura	Proturans
Psocoptera	Booklice
Siphonaptera	Fleas
Siphunculata	Sucking lice
Strepsiptera	Stylopids
Thysanoptera	Thrips
Thysanura	Bristletails, silverfish
Trichoptera	Caddis flies
Zoraptera	

48 Crowson, R.A. 1970. *Classification and biology*. 350pp. Heinemann; London.

This work adopts a strictly phylogenetic approach to the subject. Zoological, botanical and palaeontological classification is considered, but microbial classification is touched on only briefly.

49 Fernald, H.T. 1939. On type nomenclature. *Annals of the Entomological Society of America* **32**: 689–702.

A most useful glossary for the nomenclature of types.

50 Hawkes, J.G. (Ed.). 1968. *Chemotaxonomy and serotaxonomy*. 299pp. Academic Press; London.

A most useful book for some of the newer techniques now being used in taxonomy.

51 Hennig, W. 1966. *Phylogenetic systematics*. Translated by D.D. Davis and R. Zangerl. vii + 263pp. University of Illinois Press; Urbana, Illinois.

A very detailed account of this most complex subject, originally published in German as *Grundzüge einer Theorie der phylogenetischen Systematik* (Deutschen Entomologischen Institut, Berlin, 1950). The translation has made it available to a much wider audience.

52 Heywood, V.H. and **Clark, R.B.** (Eds). 1982. *Taxonomy in Europe. Final report of the Committee of European Science Research Council's Ad Hoc Group on Biological Recordings, Systematics and Taxonomy*. 170pp. North-Holland; Amsterdam. (European Science Research Council Review No. 17)

Reviews the recommendations made in the interim report in 1977 for both botany and zoology. Surveys the taxonomic resources country by country with some emphasis on tropical taxonomy. There is also some discussion on the role of and problems faced by amateur taxonomists.

53 International Commission for Zoological Nomenclature. *Bulletin of Zoological Nomenclature*. 1943 –. En. 4 per year. International Trust for Zoological Nomenclature, c/o British Museum (Natural History), Cromwell Road, London SW7 5DB, England.

Publishes cases put to the Commission relating to zoological nomenclature, and its rulings.

International code of zoological nomenclature. 1964. 176pp. International Trust for Zoological Nomenclature, London. Amendments: *Bulletin of Zoological Nomenclature* **31**: 70–101.

The *Code* is 'the set of criteria to be met in giving to an animal, or to a taxonomic group of animals, a scientific name, with its proper reference of author and date. . .'. A new revised 3rd edition is in active preparation.

Official list of family-group names in zoology
1958	38pp.	Names 1–236
1966	pp. 39–68	Names 237–282

Official index of rejected and invalid family-group names in zoology
1958	38pp.	Names 1–273
1966	pp. 39–61	Names 274–411

Official list of generic names in zoology
1958	200pp.	Names 1–1274
1966	pp. 201–267	Names 1275–1651

Official index of rejected and invalid generic names in zoology
1958	132pp.	Names 1–1169
1966	pp. 133–193	Names 1170–1743

Official list of specific names in zoology
1958	206pp.	Names 1–1525
1966	pp. 207–286	Names 1526–2045

Official index of rejected and invalid specific names in zoology
    1958   73pp.   Names 1–527
    1966   pp. 74–110   Names 528–807
Official list of works approved as available for zoological nomenclature
    1958   12pp.   Names 1–38
Official index of rejected and invalid works in zoological nomenclature
    1958   14pp.   Names 1–58

These volumes make up a complete index of the names and works that have been ruled upon by the International Commission for Zoological Nomenclature up to the time of their publication.

54 Jeffrey, C. 1973. *Biological Nomenclature*. 69pp. Systematics Association; London.

A simple guide to the principles of nomenclature. On pp. 53–69 is a glossary of the terms used in nomenclature.

55 Mayr, E. 1969. *Principles of systematic zoology*. xvi + 428pp. McGraw-Hill; New York

This work, a complete revision of the earlier edition by Mayr, Linsley and Usinger (entry 56), has become accepted as the best text on the subject of systematic zoology. It is a teaching and reference work, and includes a valuable chapter on the preparation of taxonomic papers.

56 Mayr, E., Linsley, E.G. and **Usinger, R.L.** 1953. *Methods and principles of systematic zoology*. vii + 328pp. McGraw-Hill; New York.

A first edition of Mayr (1969). The later edition has been greatly revised and enlarged, but in its absence, this edition is still a very useful textbook.

57 Mayr, E., Linsley, E.G. and **Usinger, R.L.** 1979. *Sistematik zooloojinim prensipleri*. Translated by N. Lodos. vii + 366pp. Ege Universitese Matbaasi, Izmir.

A translation into Turkish of entry 56.

58 Shiraki, T. [1961]. *Classification of insects*. [In Japanese]. 961pp. Hokuryan; Tokyo.

Although the main text of the volume is in Japanese it does give Latin scientific names. The volume contains many, but rather poorly reproduced, illustrations. It has an index to genera and species and to all English names used in the text.

59 Simpson, G.G. 1967. *Principles of animal taxonomy*. 247pp. Columbia University Press; New York.

A good standard textbook on the subject.

60 Sneath, P.H.A. and **Sokal, R.R.** 1973. *Numerical taxonomy: the principles and practice of numerical classification*. 573pp. W.H. Freeman; San Francisco.

The standard text on numerical taxonomy; the work has an excellent bibliography and is well indexed.

61 **Van der Hammen, L.** 1981. Type-concept, higher classification and evolution. *Acta Biotheoretica* **30**: 3–48.

62 **White, M.J.D.** 1978. *Modes of speciation*. 455pp. W.H. Freeman; San Francisco.

63 **Wiley, E.O.** 1981. *Phylogenetics: the theory and practice of phylogenetic systematics*. 439pp. John Wiley; Chichester.

'This book is about systematics and how the results of systematic research can be applied to studying patterns and processes of evolution.' Useful chapters on publication and the rules of nomenclature are included. There is a fairly comprehensive bibliography.

NOMENCLATORS AND CATALOGUES

Name-compendia (nomenclators) for the very early literature were completed up to 1850 and for all groups of animals by one man, C.D. Sherborn. With the great proliferation of literature in the natural sciences and greater study of the world's fauna, this kind of work could no longer be continued. However, many families of insects have been catalogued, and indeed many such catalogues are still being prepared and published. The major catalogues are included here.

Nomenclators

64 **Neave, S.A.** 1939–40. *Nomenclator zoologicus. a list of the names of genera and subgenera in zoology from the 10th edition of Linnaeus, 1758, to the end of 1935.* 4 volumes plus a supplement. Zoological Society of London; London.

1950	Volume 5	Names listed for 1936–45
1960	Volume 6	Names listed for 1946–55
1975	Volume 7	Names listed for 1956–65

Volume 8 is in preparation. However, names for new genus-group taxa published after 1965 can be traced in the last part of the annual volume of the *Zoological Record*. Section 20 lists all the new generic and subgeneric names described during the year of the volume and recorded in all the sections of the *Zoological Record*.

65 **Sherborn, C.D.** 1902. *Index animalium 1758–1800.* lix + 1195pp. Cambridge University Press; Cambridge, England.

'To provide zoologists with a complete list of all the generic and specific names that have been applied by authors to animals since 1758; to give an exact date for each page quotation and to give quotation for each reference sufficiently exact to be intelligible to both the specialist and to the layman.' Thus wrote Sherborn in the introduction to his work; these very detailed specifications were exactly adhered to and the work is most complete. On pp. xi–lix is a complete bibliography of the literature used during the compilation of the work.

66 Sherborn, C.D. 1922–33. *Index animalium 1801–1850.* 32 parts, A–Z. 7056pp. Additions and index 1096pp. British Museum (Natural History); London.

A continuation of the previous entry. The work has not been continued beyond this date, however, various groups of insects have been catalogued in a similar fashion.

Catalogues

COLEOPTERA

67 Schenkling, S. (Ed.). 1910–. *Coleopterorum catalogus.* Dr W. Junk; The Hague.

Contains original references to described names; references to biology and illustrations. Each part is separately paginated and by different authors. Parts and supplements still being published.

DERMAPTERA

68 Sakai, S. 1970–76. *Dermapterorum catalogus praeliminaris.* 1. *Bulletin Daito Bunka University* **3**: 1–49. 1a. *Special Bulletin Daito Bunka University* **1**: 1–9. 2. *Special Bulletin Daito Bunka University* **2**: 1–177. 3–6. (1) *Bulletin Daito Bunka University* **4**: 1–68. (2) *Bulletin Daito Bunka University* **1971**: 1–14. (3) *Special Bulletin Daito Bunka University* **3**: 1–62. (4) *Special Bulletin Daito Bunka University* **5**: 1–265. 7. *Special Bulletin Daito Bunka University* **6**: 1–357. 8. *Bulletin Daito Bunka University* **13**: 1–47. 9. *Special Bulletin Daito Bunka University* **13**: 1–20

DIPTERA

69 Knight, K.L. and **Stone, A.** 1977. *A catalog of the mosquitoes of the world, Diptera Culicidae.* 2nd edn. xi + 611pp. Entomological Society of America; College Park, Maryland.

70 Knight, K.L. and **Stone, A.** 1978. *Supplement to 'A catalog of the mosquitoes of the world, Diptera Culicidae'.* 2nd edn. 107pp. Entomological Society of America; College Park, Maryland.

71 Crosskey, R.W. *et al.* (Eds). 1980. *Catalogue of the Diptera of the Afrotropical region.* 1437pp. British Museum (Natural History); London.

In addition to the main catalogue entries, the introduction to each family contains much biological information and information on distribution. There is a full bibliography of the papers cited in the main body of the work and a full list of authors' names and lifespan dates with biographical references.

72 Stone, A. *et al.* (Eds). 1965. A catalog of the Diptera of America north of Mexico. *Agriculture Handbook, United States Department of Agriculture* No. **276**: 1–1696.
Corrections: *Bulletin of the Entomological Society of America* **13**: 115–25; **17**: 83–8; **24**: 143–4.

This volume also contains a full bibliography of cited works from the main text.

73 **Papavero, N.** (Ed.). 1966–. *A catalogue of the Diptera of the Americas south of the United States.* Departamento de Zoologia, Secretaria da Agricultura do Estado de São Paulo; São Paulo.

Each fascicle published is written by a separate author and has its own bibliography. Fascicles for 80 of the intended 110 families have now been completed and published.

74 **Delfinado, M.D.** and **Hardy, D.E.** (Eds). 1973–77. *A catalog of the Diptera of the Oriental region.* 3 volumes. University Press of Hawaii; Honolulu.

A full catalogue covering all the Diptera families of the Oriental region, giving the distribution and full citations for original descriptions.

HEMIPTERA–HOMOPTERA

75 *General catalogue of the Hemiptera (later Homoptera).* 1927–. Editors: Fasc. 1, W.D. Funkhouser (1927); Fasc. 2, G. Horváth (1929); Fasc. 3; R.F. Hussey and E. Shermann (1929); Fasc. 4, Z.P. Metcalf (after pt. 11 W.E. China) (1932–62). North Carolina State College; Raleigh, North Carolina.

The work gives all references to original descriptions, synonymies, etc. Several volumes are complete bibliographies to some of the families. Some of the references are inaccurate or badly quoted, making them difficult to interpret, but the work is very complete.

75A **Mound, L.A.** and **Halsey, S.H.** 1978. *Whitefly of the world.* 340 pp. British Museum (Natural History); London.

HYMENOPTERA

76 *Hymenopterorum catalogus.* 1965–. Dr W. Junk; The Hague.

There is a separate author for each family published. Parts are issued at irregular intervals, each with its own index.

77 **Krombein, K.V.** *et al.* 1979–80. *Catalogue of Hymenoptera in America north of Mexico.* 3 volumes. Smithsonian Institution Press; Washington, DC.

As with most similar catalogues, the references are arranged under families. All names have full literature references. Volume 3 is an index volume and includes indexes to Hymenoptera, Hosts, Parasites, Prey, Pollen and nectar sources and Predators.

ISOPTERA

78 **Snyder, T.E.** 1949. Catalogue of the termites (Isoptera) of the world. *Smithsonian Miscellaneous Collections* **112**: 1–490.

The catalogue is arranged systematically; known distribution is included. A full bibliography is included on living and fossil termites.

LEPIDOPTERA

79   Aurivillius, C. (Ed.). 1911–39. *Lepidopterorum catalogus* Pts 1–94. Dr W. Junk; Berlin.

Lists family-group, generic-group and species-group names; the original references, illustrations and biology references are also included.

ORTHOPTERA

80   Beier, M. (Ed.). 1938–. *Orthopterorum catalogus*. Dr W. Junk; The Hague.

Individual authors for each part issued. The work gives full literature references for all the names included, along with any other pertinent references to biology and distribution.

THYSANOPTERA

81   Jacot-Guillarmod, C.F. 1970–5. Catalogue of the Thysanoptera of the world. *Annals of the Cape Province Museum* **7**: 1–216, 217–515, 517–976, 977–1255, 1257–1556.

A complete catalogue of all the Thysanoptera names, with distribution and full literature references.

TRICHOPTERA

82   Fischer, F.O.J. 1960–73. *Trichopterorum catalogus*. 15 volumes plus an index volume. Nederlandse Entomologische Vereniging; Amsterdam.

A complete catalogue of Trichoptera names with all original references.

IDENTIFICATION AND FAUNISTICS

Papers and books on the identification of insects from various parts of the world are bewildering in their number. It is difficult enough for the specialist research worker to find his way around the literature, but for the uninitiated it is an almost impossible task. The problem has been partly overcome by the publication in the last few years of some very good guides. These include references to the more important works that have been published, both as books and as papers in scientific periodicals.

Hollis (1980) is the most useful work and is international and reliable in its coverage. Two other major groups of reference works have been included here. Many of the references will be found in Hollis (1980) or Kerrich, Hawksworth and Sims (1978), but we have thought it pertinent and helpful to the user to have them listed here.

Monographic treatments of the insects of particular countries or geographic regions ('Faunas') are still much in vogue. Whilst many old fauna series have been discontinued, others have recently started publication and their numbers are increasing. Whilst varying much in quality and scope, at their best they are invaluable identification works, and often include a wealth of data on distribution, biology, food plants, hosts, etc. Similar information can sometimes be obtained from the major catalogues (*see* the section 'Nomenclators and catalogues', p 21) of insect orders or families.

Amateur entomology is much more strongly pursued as a pastime (even as an obsessive hobby) in Britain than anywhere else in the world. We offer no apology for providing much more detailed faunistic coverage for Britain than other countries. The relative richness of the British faunistic literature to a large extent reflects the amateur interest in local insects, and the amateur student, in our experience, regularly seeks the references included.

83 Brues, C.T., Melander, A.L. and Carpenter, F.M.   1954.   Classification of insects. Keys to the living and extinct families of insects and to the living families of other terrestrial arthropods. *Bulletin of the Museum of Comparative Zoology* **108**: 1–907.

Identification keys to the taxonomic level of family and subfamily. The work contains for each order a geographically arranged bibliography of essential taxonomic literature.

84 D'Abrera, B.   1977–. *Butterflies of the world*; Vol. 1 (1977), *Butterflies of the Australian region* (415pp); Vol. 2 (1980), *Butterflies of the Afrotropical region* (593pp); Vol. 3 (1981), *Butterflies of the Neotropical region* (in four parts); Vol. 4 (in preparation), *Butterflies of the Oriental region*; Vol. 5 (in preparation), *Butterflies of the Holarctic region*. Lansdowne Editions; Melbourne / E.W. Classey; Faringdon, Oxfordshire, England.

The series aims to cover the whole of the world's butterfly fauna, and is aimed at collectors, students and research institutions. The volumes are in a large format, and very fully illustrated with often superb colour photographs.

85 Hollis, D. (Ed.). 1980. *Animal identification: a reference guide*; Vol. 3, *Insects*. 160pp. British Museum (Natural History) and John Wiley; London and Chichester.

This work may be regarded as *the* guide to identification keys. 'The main objective of the volume is to provide a list of primary references which will enable the non-specialist to identify insects from any part of the world.' The work is divided by order of insects, and then by major biogeographical regions. An easy-to-use volume and well indexed.

86 Kerrich, G.J., Hawksworth, D.L. and Sims, R.W.   1978.   Key works to the fauna and flora of the British Isles and north-western Europe. *Systematics Association Special Volume* No. **9**. xii + 179pp.

Pages 58–110 of this volume cover the insect families. The work is a good guide to the most important works on European insects; it covers more general works in addition to papers for identifying insects.

87 Peterson, A. 1948. *Larvae of insects: an introduction to Nearctic species.* 2 volumes. Edwards; Columbus, Ohio.

Although the work covers instructions for collecting and preserving larvae, and the importance of larva recognition, it is essentially a work for the identification of larvae. It is a well illustrated work with good line drawings, showing the essential features for identification. Part 1 covers the Lepidoptera and Hymenoptera; part 2 the Coleoptera, Diptera, Neuroptera, Siphonaptera, Mecoptera and Trichoptera.

88 Scheiding, U. Göllner-. 1967–71. Bibliographie der Bestimmungstabellen europäische Insekten (1880–1963).
Tl 1, Apterygota bis Siphonaptera. *Beiträge zur Entomologie* **17**: 697–958.
Tl 2, Hymenoptera. *Mitteilungen aus dem Zoologischen Museum in Berlin* **45**: 1–156.
Tl 3, Coleoptera und Strepsiptera. *Deutsche Entomologische Zeitschrift* **17**: 33–118, 433–76; **18**: 1–84, 287–360.

A very full bibliography of references to identification works for European insects.

Faunas

AFRICA (*see also* Madagascar)

89 Skaife, S.H. 1979. *African insect life.* 2nd edn, revised by J. Ledger. 279pp. Country Life Books; London.

The new edition is in an entirely different format from the earlier and much used edition. It covers the insect fauna of Africa south of the Sahara in a semipopular but very thorough fashion. Many black-and-white illustrations plus 72 superb colour plates. Useful as a general text on entomology for the lay reader and student.

BRAZIL

90 da Costa Luna, A. (Ed.). *Insetos do Brasil.* 12 volumes 1938–62. Escuola Nacional de Agronomia; Rio de Janeiro.

The families of insects are covered in individual volumes. Keys and descriptions are included. The volumes are well illustrated.

CANADA (*see also* entries 114–118)

91 *The insects and arachnids of Canada.* 8 volumes 1977–81. Agriculture Canada; Ottawa.

Each volume has a separate author. The volumes are written to assist student, agriculturalist and forestry worker to study the insects of Canada. Species are described; data are given on the known habitats and host plants are given where applicable. Volume 1 deals with the collecting and preserving of specimens.

CZECHOSLOVAKIA

92 *Fauna CSSR*. 1954–. Verlag der Tschechoslowakischen Akademie der Wissenschaften; Prague.

This is a general fauna of Czechoslovakia, but many of the volumes so far issued cover insect families. Volumes appear at irregular intervals. The language is mostly Czech, but some parts are written in German.

93 *Klic zvireny CSSR* [Keys to the fauna of Czechoslovakia]. Česko-slovenská Akademie Věd; Prague.

All volumes are written in Czech, each with a separate author and covering a family of insects. The volumes are published at irregular intervals.

DENMARK

94 *Danmarks Fauna*. 1907–. Dansk Naturhistorisk Forening; Copenhagen.

Each volume is devoted to a particular family of insects. The series is written in Danish, and is a complete guide to the families, with descriptions of species, descriptions of habitats and localities; some volumes are illustrated.

FRANCE

95 *Faune de France*. 1922–64. Fédération Française des Sociétés de Sciences Naturelles; Paris.

Forty volumes of this general fauna cover the insect families. Individual families of insects are allocated to separate volumes written by specialist authors. They were issued at irregular intervals and are in French. Species are all described including new taxa; the work also carried details of localities and habitat. Some volumes are illustrated.

GERMANY

96 *Die Tierwelt Deutschlands. Insecta*. 1925–. VEB Gustav Fischer Verlag; Jena.

A long series, issued at irregular intervals. The series aims to cover the general fauna of Germany. Many of the volumes are on insects and several families of insects have now been described, in some detail.

HAWAII

97 *Insects of Hawaii*. 1948–. University of Hawaii Press; Honolulu, Hawaii.

A scholarly work, 13 volumes of which have now been issued. Each family is fully described by its own specialist author. Volumes are issued at irregular intervals.

HUNGARY

98 *Fauna Hungariae*. 1955–. Akadémiai Kiadó; Budapest.

The insect parts of this series are all written in Hungarian, by different authors, appearing frequently but at irregular intervals. Each part, covering one family, is complete in itself, with its own index. The series has a dual numbering system, making for some considerable confusion when quoting the parts. Care should be taken when quoting references in bibliographies; and it is recommended that the general *Fauna Hungariae* numbers should be used – this does make for ease of retrieval.

IRELAND – *see* entries 121–138

ITALY

99 *Fauna d'Italia*. 1965–. Accademia Nazionale Italiani di Entomologia e dell'Unione Zoologica Italiana. Edizioni Calderini; Bologna.

A series which aims eventually to cover the entire fauna of the country. Several insect families have been covered to date. It is written in Italian; each family by a specialist author.

JAPAN

100 *Fauna Japonica*. 1960–. Keigaku; Tokyo.

Written in English by specialist authors, the volumes are well illustrated. They are issued at frequent but irregular intervals, and many that have appeared cover insect families.

MADAGASCAR

101 **Paulian, R.** (Ed.). *Faune de Madagascar*. 1956–. Centre National de la Recherche Scientifique; Paris.

Most of the parts issued are written in French, but a few are in English. Keys for identification are given; descriptions of taxa and the full distribution and biology are given for each species.

MICRONESIA

102 *Insects of Micronesia*. 1954–. Bernice P. Bishop Museum; Honolulu, Hawaii.

Each volume is given over to a particular insect family, but volumes are issued in parts. Many of the volumes are still incomplete, but parts are still being issued at irregular intervals. Written in English by specialist authors; poorly illustrated.

NEW ZEALAND

103 *Fauna of New Zealand*. 1982–. DSIR; Wellington.

Each contribution will cover a major taxonomic unit such as a family.

POLAND

104 *Klucze do oznaczania owadow polski* [Keys for the identification of Polish insects]. Państwowe Wydawnictwo Naukowe; Warsaw.

Issued at irregular intervals. The series has a complex numbering system, but basically each volume covers an insect family; numerous parts make up the volume, each written in Polish by a specialist author on a particular group. The series give detailed keys for the identification to species level.

ROMANIA

105 *Fauna Republicii Socialiste România. Insecta.* 1951–. Academia Republicii Socialiste România; Bucharest.

The series is written in Romanian, each part by a specialist author. The work includes keys for identification to species level, descriptions of new taxa and information on localities and biology.

SAUDI ARABIA

106 **Wittmer, W.** and **Büttiker, W.** (Eds). *Fauna of Saudi Arabia.* Pro Entomologia, c/o Natural History Museum, Basle and Ciba-Geigy, Basle.

This series now has four volumes published and a fifth volume due to be issued in 1983. About 70 per cent of the text of the three volumes issued is devoted to insect families. The volumes are well illustrated.

SCANDINAVIA

107 *Fauna entomologica scandinavica.* 1973–. Scandinavian Science Press; Gadstrip, Denmark.

Issued as separate volumes at irregular intervals. The treatment is at generic and specific levels. It includes both local and total distribution, habitats and biology. The series includes the fauna of north Germany and the UK to North Cape and East Karelia.

SOUTH AMERICA

108 **Hurlbert, S.H.** (Ed.). 1977. *Aquatic biota of southern South America.* 342pp. San Diego State University; San Diego.

A compilation of taxonomic bibliographies for the fauna and flora of inland waters of southern South America.

109 **Hurlbert, S.H., Rodriguez, G.** and **Dias dos Santos, N.** (Eds). 1981. *Aquatic biota of tropical South America.* 2 parts. San Diego State University; San Diego.

Compilation of taxonomic bibliographies for the fauna and flora of inland waters of the tropical portion of South America.

SPAIN

110 Gomez-Bustillo, M.R. and **Fernández-Rubio, F.** 1974–80. *Mariposas de la Peninsula Ibérica.* 4 volumes. Instituto Nacional par la Conservación de la Naturaleza; Madrid.

Aims to cover the moth and butterfly fauna of Spain.

SWITZERLAND

111 *Insecta Helvetica catalogus.* 1966–. Société Entomologique Suisse; Lausanne.

Very few volumes have been issued to date; they are issued at irregular intervals.

112 *Insecta Helvetica fauna.* 1964–. Société Entomologique Suisse, Lausanne.

Only four volumes have so far been issued, three on the Hymenoptera and one on Coleoptera.

UK – *see* entries 121–138

USA (*see also* Hawaii)

113 *Arthropods of Florida.* 1965–. Florida Department of Agriculture; Gainesville, Florida.

This series is still appearing. The insect parts are as follows: Pt 1, *Lepidoptera. A checklist*; Pt 3, *Armored scale insects*; Pt 7, *Agromyzidae*; Pt 8, *Scarab beetles*; Pt 9, *Ichneumonidae*; Pt 10, *Sandflies*. Each of the parts, with the exception of the Lepidoptera checklist, is a detailed study of the group.

114 Darsie, R.F. and **Ward, R.A.** 1981. Identification and geographical distribution of the mosquitoes of North America, north of Mexico. *Mosquito Systematics, Suppl.* **1** : 1–313.

Illustrated keys for the identification of larvae and adults are given. Distribution maps are given for mosquito taxa.

115 Ferguson, D.C. and **Dominick, R.B.** 1971–. *Moths of America north of Mexico, including Greenland.* E.W. Classey & RBD Publications; Faringdon, Oxfordshire, England.

Published at irregular intervals, as each part is completed (not in numerical order).

116 McAlpine, J.F. *et al.* (Eds). 1981. *Manual of Nearctic Diptera*, vol. 1. Research Branch, Agriculture Canada. (Monograph No. 27.)

The work aims to provide an up-to-date, well-illustrated means of identifying flies of America north of Mexico. The second (and final) volume is yet to appear. Suitable for professional biologists, teachers, students and well-informed amateurs.

117   Merritt, R.W. and **Cummins, R.W.** (Eds). 1978. *An introduction to the aquatic insects of North America.* viii + 441pp. Kendall/Hunt; Dubuque, Iowa.

The first six chapters deal with the general morphology, ecology and distribution and collecting and sampling; these introductory chapters are then followed by more descriptive chapters on each of the aquatic families of insects. The illustrations are excellent; the volume carries a very full bibliography of some 1712 references and is well indexed.

118   Swan, L.A. and **Papp, C.S.** 1972. *The common insects of North America.* xiii + 750pp. Harper & Row; New York.

A useful survey of the more common insects of the USA. Details of distribution and food plants are given. There is an index of scientific names, subject and common names, as well as an appendix covering geological eras, the place of insects in time and the orders and families represented in the book. On pp. 678–92 is a glossary of terms used in the volume.

USSR

119   *Fauna SSSR.* 1935–. Zoological Institute, Academy of Sciences of the USSR; Leningrad.

Volumes of this long-established fauna are still being issued and updated new editions of some of the earlier issues are appearing. Each volume has its own specialist author. The whole series is written in Russian, but some of the volumes have now been issued in cover-to-cover translated editions – most of them by the Smithsonian Institution in Washington DC. Some of the earlier translations were issued by the Israel Program for Scientific Translation. Parts of some volumes, particularly the identification keys, have been translated through the British Library Lending Division's translation programme.

120   *Fauna Ukraine.* Akademiya Nauk Ukrainskoi SSR. Institut Zoologii; Kiev.

Each volume is a complete descriptive treatment of the insect families so far covered. Some new taxa are described and the distribution as well as the biological data are given for each species.

Insect fauna of the British Isles

There are three main organizations which produce some excellent handbooks, useful for the amateur and professional alike. They are very full in their treatment of the subjects, and are well produced and moderately priced.

121   Amateur Entomologists' Society

Details of the serial publications of this Society will be found in the 'Primary journals, review journals, monograph series' section. Two of their regular series have included a number of specialist handbooks:

Amateur Entomologist No. 6 (1982), *Silkmoth rearer's handbook* (3rd edn); No. 7 (1982), *Hymenopterist's handbook* (160pp) (reprinted);

No. 11 (1954), *Coleopterist's handbook* (120pp); No. 13 (1976), *Lepidopterist's handbook* (138pp); No. 4 (1976), *Insect photography for the amateur* (52pp); No. 5 (1978), *A dipterist's handbook.*
Leaflets: *Rearing and studying the praying mantid; Study of the genitalia of Lepidoptera; Insect light traps.*

This is just a selection of the many titles available.

122   Freshwater Biological Association

This body has issued a numbered series of publications, several of which deal with aquatic entomology. Among the many titles included are: *Simuliidae* (black-flies); *Plecoptera* (stoneflies); and *Trichoptera* (caddis flies) They usually cover both the adults and the larvae and include identification keys.

123   Royal Entomological Society of London
Handbooks for the identification of British insects. 1954–.

A series of volumes issued in parts, aiming eventually to cover all groups of British insects; besides providing identification keys, there is also a great deal of information on biology and distribution.

The following books are useful guides to the British insect fauna.

124   Brooks, M. and **Knight, C.** 1981. *A complete guide to British butterflies.* 159pp. Jonathan Cape; London.

Unusually, the life histories of species are shown by photograph rather than descriptively in this volume. The distribution, habitat, life cycle and larval food plants are given for all species. A list of scientific and common names is included. There is a useful chapter on collecting, breeding and photography. Useful to both the amateur and the professional.

125   Carter, D. 1982. *Butterflies and moths in Britain and Europe.* 192pp. British Museum (Natural History); London.

A useful general book for popular reading. It has some excellent colour photographs. The text is brief but covers a range of topics such as plants to attract butterflies to the garden and the protection and conservation of species.

126   Carter, D. 1979. *The observers' book of caterpillars.* 159pp. Frederick Warne; London.

127   Colyer, C.N. and **Hammond, C.O.** 1968. *Flies of the British Isles.* 2nd edn. 384pp. Frederick Warne; London.

128   Corbet, P.S. and **Longfield, C.** 1960. *Dragonflies.* xii + 260pp. Collins; London. (New Naturalist Series No. 41.)

129   Ford, E.B. 1945. *Butterflies.* xiv + 368pp. Collins; London. (New Naturalist Series No. 1.)

130   Ford, E.B. 1972. *Moths.* 3rd edn. xix + 266pp. Collins; London. (New Naturalist Series No. 30.)

131 **Ford, R.L.E.** 1973. *Studying insects.* xii + 150pp. Frederick Warne; London.

132 **Hammond, C.O.** 1977. *The dragonflies of Great Britain and Ireland.* 115pp. Curwen Press; London.

133 **Heath, J.** and **Emmet, A.M.** (Eds). 1976–. *The moths and butterflies of Great Britain and Ireland.* 11 volumes (2 published to date). Curwen Press; London.

An ambitious series, aiming to cover the British Lepidoptera in quite some detail, incorporating life history and distribution data for most species. Each family is dealt with by an expert author, and the series is extremely well illustrated with colour drawings, colour photographs and line drawings as an aid to identification. It is unfortunate that publication has slipped so far behind schedule, although the two volumes so far published can be used on their own. Volume 11 is to be devoted to larvae, illustrated in colour.

134 **Higgins, L.G.** and **Riley, N.D.** 1980. *A field guide to the butterflies of Britain and Europe.* 4th edn. 384pp. Collins; London.

This work has also been translated into the following languages: Danish, Dutch, Finnish, French, German, Norwegian, Spanish and Swedish. An American edition has also been produced.

135 **Howarth, T.G.** 1973. *South's British butterflies.* 210pp. Frederick Warne; London.

The plates are reproduced in *Colour identification guide to British butterflies*, published as an accompanying volume.

136 **Linssen, E.T.** 1959. *Beetles of the British Isles.* 2 volumes. Frederick Warne; London.

137 **Ragge, D.R.** 1965. *Grasshoppers, crickets and cockroaches of the British Isles.* xii + 299pp. Frederick Warne; London.

138 **Southwood, T.R.E.** and **Leston, D.** 1959. *Land and water bugs of the British Isles.* x + 436pp. Frederick Warne; London.

High-altitude, Arctic and Subantarctic insects

139 **Danks, H.V.** 1981. *Arctic arthropods: a review of systematics and ecology with particular reference to the North American fauna.* 608pp. Entomological Society of Canada; Ottawa.

The work is divided into three parts: features of the Northlands, the arthropod fauna and its relationships, and an annotated list of reported North American Arctic species, and their known distribution. There are some 2015 references.

140 **Danks, H.V.** 1981. *Bibliography of the Arctic arthropods of the Nearctic region.* 123pp. Entomological Society of Canada; Ottawa.

A bibliography of 1382 references arranged alphabetically. Further indexes are provided: an index to subsidiary authors, a geographic index, a taxonomic index and a subject index. A most useful and up-to-date bibliography.

141 **Downes, J.A.** 1964. Arctic insects and their environment. *Canadian Entomologist* **96**: 279–307.

The paper discusses the characteristics of the Arctic environment as they influence insect life; the make-up of the Arctic fauna, and a discussion of the reasons why so few species are found in the Arctic.

142 **Downes, J.A.** 1965. Adaptations of insects in the Arctic. *Annual Review of Entomology* **10**: 257–74.

Physiological and ecological adaptations are discussed. The paper has a good bibliography.

143 **Gressitt, J.L.** 1967. *Entomology of Antarctica.* xii + 375pp. American Geophysical Union; Washington, DC. (Antarctic Research Series No. 10.)

A short introduction gives some historical background to entomological discoveries in Antarctica. Most of the entomological fauna is treated and most of the species keyed.

144 **Gressitt, J.L.** 1970. Subantarctic entomology, particularly of South Georgia and Heard Island. 374pp. *Pacific Insects Monograph* **23**: 1–374.

The introductory chapters discuss the entomological exploration of the two main areas, and give some history of general exploration of the sites. Arachnida are included in addition to the Insecta.

145 **Gressitt, J.L.** and **Weber, N.A.** 1959. Bibliographic introduction to Antarctic–Subantarctic entomology. *Pacific Insects* **1**: 441–80.

A background and reference point for zoogeographic studies of the area. An annotated bibliography of 424 items is included. Nothing similar is available.

146 **Mani, M.S.** 1962. *Introduction to high-altitude entomology: insect life above the timber line in the north-west Himalaya.* xix + 302pp. Methuen; London.

The ecology and zoogeography of the area are discussed; the origins and evolution of the nival insect fauna, their ecologic specializations and interrelations of particular groups of insects are considered in some detail.

147 **Mani, M.S.** 1968. Ecology and biogeography of high-altitude insects. *Series Entomologica* No. **4**: 1–527.

Subantarctic and Arctic insects, their ecology and distribution are covered in Chapters 15 and 16 of this work. The remainder of the work is devoted to mountain regions of the world and their associated insect faunas. It has a very full bibliography.

Distribution maps

BELGIUM (the European Invertebrate Survey)

148 **Leclercq, J.** (Ed.). 1970–. *Atlas provisoire des insectes de Belgique.* Maps 1–. Faculté des Sciences Agronomiques de l'Etat Zoologie Générale et Faunistique Gembloux.

BRITISH ISLES (the European Invertebrate Survey)

149 **Heath, J.** (Ed.). 1970–. *Provisional atlas of the insects of the British Isles.* 1–. Biological Records Centre, Monks Wood Experimental Station; Abbots Ripton, Huntingdon, England.

150 **Lamhna, E.N.** (Ed.). 1978–. *Provisional atlas of the dragonflies of Ireland.* Irish Biological Records Centre; St Martins House, Waterloo Road, Dublin 4.

USSR (the European Invertebrate Survey)

151 **Gorodkov, K.B.** (Ed.). 1978–. *Provisional atlas of the insects of the European part of the USSR.* 1–. [In Russian]. Izdatel'stvo 'Nauka', Moscow.

INTERNATIONAL

152 **Commonwealth Institute of Entomology.** 1951–. *Distribution maps of pests.*

Eighteen new or revised maps are issued each year, each map showing the world distribution of a single agricultural insect pest. The reverse of each map lists supporting bibliographic references. Over 400 maps have been issued so far. An index is issued at irregular intervals.

KOREA

153 **Chan-Whan Kim** (Ed.). 1976–. *Distribution atlas of insects of Korea.* Volumes 1–. Korea University Press; Seoul.

IDENTIFICATION SERVICES

We have attempted to cover in this volume not just the literature but all types of information source in entomology. We deal now with one of the most important and, incidentally, most undervalued of the non-bibliographic services.

We indicated in our preface the tremendous economic importance of insects, and touched on the potential for biological control of harmful species by their parasites and predators. It is now generally accepted that great care needs to be exercised when attempting to control a pest, as it is all too easy to destroy not only the target species but also its associated natural enemies. It is therefore vital that all the species involved are accurately identified and their roles understood before any attempt is made to alter the delicate balance of the ecosystem.

Because of the number and diversity of species, insect identification is a highly specialized field, the province of a relatively small number of taxonomists who research and generally concern themselves with just one group of insects. This degree of specialization can normally be supported only by museums, university research departments, government agencies and some of the professional organizations. Many of these can provide limited identification services for outsiders, but such services are usually incidental to the main functions of the organizations.

Freelance or amateur taxonomists, together with some professionals, can be located through the 'identifiers' index of R.H. Arnett's *The naturalists' directory and almanac (international)* (43rd edition published in 1978 by World Natural History Publications, Baltimore, USA); there is of course no guarantee as to the competence or authority of all those listed.

Reference may also be made to the American Registry of Professional Entomologists (ARPE), which can help in locating properly qualified professionals. Details of ARPE are included under the heading 'Professional organizations'.

There is however one organization which provides an authoritative and international service in this field: the Commonwealth Institute of Entomology (CIE). The CIE team of taxonomists is based in the British Museum (Natural History), which houses the UK National Insect Collection, a unique assemblage of over 22 million specimens representing some 450 000 species from around the world. Using this remarkable collection as a reference tool, the CIE's taxonomists identify over 50 000 insects and mites each year. This service, which is limited to agriculturally important species (harmful or beneficial) is, at the time of writing, open to all. It is free to most types of user in countries that are members of the Commonwealth Agricultural Bureaux, but a charge is made for international or commercial organizations and for users in non-member countries. Enquiries concerning the CIE service should be addressed to:

154   Commonwealth Institute of Entomology. 56 Queen's Gate, London SW7 5JR, England (for attention of the Director).

COMMON NAMES OF INSECTS

The citing of common names without their associated scientific name is not unusual, particularly in the popular literature; but their use is not exclusive to this type of publication. Common names are often used quite liberally throughout economic and report literature.

Their use can be misleading to research workers. The same or a similar common name can be used for very different insects in different parts of the world and indeed sometimes in different parts of the same country. It is important, therefore, to research workers and others to know exactly what species is really in question. As an example, T.K. Crosby in *New Zealand Entomologist* **5** : 336–9 (1973) cites the case of the use of the terms 'sandfly' and 'black fly' in New Zealand. Black-fly is the usual name in many parts of the world used for flies of the family Simuliidae (blackfly as one word is a common name for some aphids); however, in New Zealand the common name is sandfly. The use of the common name sandfly in most other regions usually refers to the subfamily Phlebotominae. Countless numbers of similar instances could be referred to. Fortunately many lists of common names with their Latin equivalents have been compiled and published, and for many parts of the world.

AFRICA AND ASSOCIATED ISLANDS (*see also* East Africa)

155 Agence de Cooperation Culturelle et Technique (Ed.). 1977. *Nomenclatures de la faune et de la flore (latin, français, anglais). Afrique au Sud du Sahara, Madagascar, Mascareignes*. 187pp. Hachette; Paris.

Only a limited number of references are included, but it has a good list of insect names; it is particularly useful for medical and veterinary insects.

ARABIC NAMES

156 Aharoni, I. 1922. *The locust, its anatomy and life-history; Arabic names and verbs relating to the locust*. 86pp. Ha-Arbeh; Jerusalem.

Although published some time ago, this list is a useful guide for names used in older literature in connection with locust plagues and crop damage.

157 FAO. 1980. *Trilingual glossary of terms used in acridology*. 171pp. Food and Agriculture Organization of the United Nations; Rome.

Listed in three languages: French, English and Arabic.

ARGENTINA

158 Lieberman, J. 1971. Sobre algunos nombres vulgares de ortópteros argentinos. *Idia* No. **283**: 78–80. Instituto Nacional de Tecnología Agropecuaria; Buenos Aires.

Not a very comprehensive list, and deals only with the Orthoptera.

159 Lizer y Trelles, C.A. 1944. *Insectos y otros enemigos de la Quinta*. 2nd edn. 217pp. (*Enciclopedia Agropecuaria Argentina*, Vol. 2.) Editorial Sudamericana; Buenos Aires.

Rather old, but includes a useful index to scientific and common names of agriculturally important insect pests.

AUSTRALIA

160 Carne, P.P., Crawford, L.D. *et al.* 1980. *Scientific and common names of insects and allied forms occurring in Australia*. iii + 95pp. CSIRO; Canberra.

In addition to lists of arthropods (and other invertebrates) under their common and scientific names, a list of scientific names in systematic order from phylum to order is provided.

BRAZIL

161 Biezanko, C.M. 1948. Nomes populares dos lepidópteros no Rio Grande do Sul. Pt 1. *Agros, Pelotas* **1**: 164–77.

162 Biezanko, C.M. 1948. Nomes populares dos homópteros no Rio Grande do Sul. *Argos, Pelotas* **1**: 249–54.

163 Biezanko, C.M. and **Link, D.** 1972. Nomes populares dos lepidópteros no Rio Grande do Sul. *Boletim técnico Universidade Federal de Santa Maria* **4**: 1–15.

CANADA

164 Benoit, P. 1975. *Noms français d'insectes au Canada avec noms latins et anglais correspondants.* 214pp. Agriculture, Quebec. (Publication No. QA 38-R4-30.) Errata published in *Annals of the Entomological Society of Quebec* **22** : 166 (1977); *Bulletin of the Entomological Society of Canada* **9**: 148–50 (1977).

An excellent, well cross-referenced work, with indexes for both Latin and common names.

CHILE

165 Brücher, G. 1942. Lista de algunos nombres vulgarés de insectos. *Boletín del Departmento de Sanidad Vegetal, Santiago de Chile* **2**: 120–5.

Listed under Latin names with common-name equivalents.

CHINA

166 Xin Jie-lu and **Xia Song-Yun.** 1978. *Glossary of common names of insects in English and Chinese.* 441pp. Hunan People's Press; Chansha.

Names listed in Latin, Chinese and English. The work covers all groups of insects with particular emphasis on pest names.

COLOMBIA

167 Apolinar Maria, H. 1937. Vocabulario de términos vulgares en historia natural Colombiana. *Revista de las Academia Colombiana de Ciencias Exactas, Bogotá* **1**: 196–206.

Mostly consists of the names of economically important plants and animals. A rather poor list for insects.

168 Posada, O.L. *et al.* 1970. Lista de insectos dañinos y otras plagas en Colombia. *Publicaciones misceláneas. Instituto Colombiana Agropec* No. **17**: ix + 202pp.

A revision of an earlier list of injurious insects and other pests in Colombia. Insects are arranged systematically under host plants, and both scientific and common names are given. There is no index.

CUBA

169 Bruner, S.C., Scaramuzza, L.C. and **Otero, A.R.** 1975. *Catálogo de los insectos que atacan a las plantas económicas de Cuba.* 2nd edn. 401pp. Academia de Ciencias de Cuba; Havana.

A revision of a catalogue of arthropods attacking economically important plants in Cuba, first published in 1945. Pests are listed under their host plants, which are arranged systematically. Spanish, and in some cases English, common names are given, and for many arthropod species an indication is given of natural enemies. There is an index by orders to the arthropods mentioned, an index to the natural enemies mentioned, and one to the common names of plants.

EAST AFRICA

170 Crowe, T.J. 1967. Common names for agricultural and forestry insects and mites of East Africa. *East African Agriculture and Forestry Journal* **33**: 55–63.

A relatively short list arranged alphabetically, first by scientific names and followed by an alphabetical list of common names with their equivalents of scientific names. Veterinary and medical insects are not included.

ESTONIA

171 Ristkok, J. (Ed.). 1977. Catalogus macrolepidopterorum Estoniae. *Abiks Loodusevaat* **72**: 1–40.

The work is basically a catalogue of the moths of Estonia, but it also includes the common names. The list is arranged alphabetically under the scientific name, with the Estonian common-name equivalent.

EUROPE (*see also* individual names of countries)

172 Gozmany, L. 1979. *Vocabularium nominum animalium Europae septem linguis redactum.* 2 volumes. Akadémiae Kiadó; Budapest.

A monumental work based on those animals (including insects) specifically inhabiting, or introduced but now established in, Europe. It covers taxa that have vernacular names in at least two of the modern languages treated, i.e. Latin, German, English, French, Hungarian, Spanish and Russian. In Vol. 1 animals are indexed under their Latin names, which are followed by a number – these are the numbers referred to in Vol. 2. This second volume is an index of the vernacular names under each language. Numbers refer the user back to the scientific name in Vol. 1. This is an excellent work covering several hundreds of names in seven languages and is easy to use despite its being split into two separate volumes.

173 Grzimek, B. 1975. *Grzimek's Animal life encyclopedia*, Vol. 2. 643pp. Van Nostrand Reinhold; New York.

Pages 565–618 list insect names in English with their equivalents in French, German and Russian.

174 Soenen, A. 1978. *Wörterbuch bedeutender Krankheiten und Schädlinge in Land- und Gartenbau (English, Español, Français, Italiano, Nederlands).* 55pp. Bayer Pflanzenschutz; Leverkusen.

The list covers the pests and diseases in agriculture and horticulture. The work is arranged in a useful tabular form, divided by insects, families and diseases. Although not actually included in the title, German names are also given. A most useful multilingual work.

FRANCE

175 Bonnemaison, L. 1962. *Les ennemis animaux des plantes cultivées et des forêts.* 3 volumes. Editions Sep; Paris.

This work, subsequently translated into Spanish, covers a large number of plant and forest pests, which are indexed under scientific and French common names.

GERMANY

176 Herfs, A. 1968. Vulgärnamen für Termiten. *Anzeiger für Schädlingskunde* **41**: 55–7.

A short but useful list of common names to this very specialized group (termites); arranged under common names.

177 Jacobs, W. and **Renner, M.** 1974. *Taschenlexicon zur Biologie der Insekten.* 635pp. Gustav Fischer; Stuttgart.

An illustrated encyclopedia of entomology including 'see' references from German common names. The text is well supported by references, and there is an extensive bibliography.

178 Schmidt, G. 1970. Die deutschen Namen wichtiger Arthropoden. *Mitteilungen aus der biologischen Bundesanstalt für Land- und Forstwirtschaft Berlin–Dahlem* **137**: 1–222.

Section 1 is an alphabetical listing of German common names for higher taxonomic categories, followed by their Latin equivalents. This order is reversed in section 2; section 3 lists common names of individual species; section 4 lists scientific names in order of their specific names. An index of generic names completes an extensive and easy-to-use work which covers more important insects and mites of all regions.

179 Weidner, H. 1971. *Bestimmungstabellen der Vorratsschädlinge und des Hausungeziefers Mitteleuropas.* viii + 223pp. Gustav Fischer; Stuttgart.

This publication contains an index of German, French and English common names on pp. 216–23.

INDIA

180 [Anon.] 1973. Insect and other pests of agricultural importance in India. *Plant Protection Bulletin, India* **22**(4) (1970) : 1–101.

Lists pests by common names and gives their scientific names and distribution. All are arranged and indexed according to host plants.

181 Vasantharaj, D.B. and **Kumaraswami, T.** 1975. *Elements of economic entomology.* vii + 508pp. Popular Book Depot; Madras.

Although the index to this volume is not very useful, common names may be found by reference to the text, which lists insects under their host plants using common names.

INTERNATIONAL

182 Merino-Rodríguez, M. 1964. *Elsevier's Lexicon of parasites and diseases in livestock.* 125pp. Elsevier; Amsterdam.

The first section is a systematically arranged list of scientific names, with their equivalent common names in English, French, Italian, Spanish and German. The work includes parasites and diseases of all farm and domestic animals, free-living wild fauna, honey bees and silkworms.

183 Merino-Rodríguez, M. 1966. *Elsevier's Lexicon of plant pests and diseases.* 351pp. Elsevier; Amsterdam.

The general arrangement is as for the above entry.

184 Pawar, A.D. 1975. Common names (including scientific names and important synonyms) and distribution of major insect pests of the rice of the world. *Rice Entomology Newsletter* No. **2**: 7–13.

This is not an exhaustive listing of common names actually in use, but a list of recommended common names that have now found wide acceptance.

185 Pflanzenschutz Anwendungstechnik der Bayer AG Leverkusen (Eds). 1976. *Verzeichnis der wichtigsten tierischen Schädlinge und Nützlinge, ihre Synonyme, Common Names und Abkürzungen.* 753pp. Publisher as editors.

Gives the names in 'all' (!) languages, listed under scientific names. The work includes an index to generic and specific names. A very comprehensive and most useful work. The languages covered are: Danish, German, English, French, Hebrew, Italian, Japanese, Dutch, Norwegian, Portuguese, Spanish, Turkish, Finnish, Swedish and Afrikaans.

IRAN

186 Gardenhire, R.Q. 1959. Summary of insect conditions in Iran. *Entomologie et Phytopathologie Applicata* **18**: 51–62.

Essentially a list of commercially important insects, but the listing is by common name with scientific equivalents. Not a very full list, but useful as no other list exists.

ITALY

187 Della Beffa, G. 1961. *Gli insetti dannosi all'agricoltura ed i moderni metodi e mezzi di lotta.* 3rd edn. xx + 1106pp. Ulrico Hoepli; Milan.

Although some of the scientific names are a little dated, the index to this volume is useful in that it includes both scientific and Italian common names for agriculturally important species.

JAPAN

188 Shiraki, T. 1952. *Catalogue of injurious insects in Japan (exclusive of animal parasites).* Preliminary study No. 71. 7 volumes. General Headquarters, Supreme Commander for Allied Powers; Tokyo.

Volumes 1–5 are a complete list of injurious insects listed under their scientific names. The work gives the Japanese vernacular name (transliterated) and the English common name, in the remaining volumes.

MEXICO

189 Peña, M.R. and **Antonio Sifuentes, J.** 1972. Lista de nombres científicos y comunes, de plagas agrícolas en México. *Agricultura Técnica en México* **3** (4): 132–44.

Lists the common and scientific names of 440 species of insects and other arthropods that are pests of the principal crops in Mexico. The list is arranged alphabetically by scientific names, but it would not be too difficult to find a particular common name by quickly scanning the appropriate column.

NETHERLANDS

190 Boersma, J. 1981. *It wylde wrimelt. List fan ynsektenammen.* 57pp. Fryske Akademy; Leeuwarden, Netherlands.

A systematically arranged list of insects known to occur in Friesland (Netherlands). The species are listed under their Friesian common names and their Latin names and Dutch common names are also given. Alphabetical lists of the Friesian, Latin and Dutch names are also provided.

NEW ZEALAND

191 Ferro, D.N. 1977. Standard names for common insects of New Zealand. *Bulletin of the Entomological Society of New Zealand* No. **2**: 1–42.

An excellent and easy-to-use list, it is indexed under both scientific and common names.

192 Miller, D. 1970. *Native insects.* 64pp. A.H. & A.W. Reed; Sydney and Auckland.

English and Maori names are used in the text and indexed on pp. 61–62.

NORWAY

193 Fjelddalen, J. *Norske dyrenavn B. Insekter of edderkoppdyr.* 77pp. Norsk Zoologiisk Forening; Oslo.

The publication includes insects and arachnids. It is indexed under Norwegian and Latin names.

PERU

194 Garcia, A. and **Garcia, R.J.** 1977. Nombre de algunas insectos y otros invertebrados en 'Quecha'. *Revista Peruana de Entomología* **19** (1976): 13–16.

An alphabetical list of arthropod common names, and related terms, in Quechua (a Peruvian Amerindian language), with Latin equivalents and explanations in Spanish.

195 Wille, T.J.E. 1952. *Entomología agricola del Peru.* 2nd edn. 543pp. Ministerio de Agricultura; Lima.

Although the text is all in Spanish, there is a useful index to scientific and common names of agriculturally important insects in Peru.

RUSSIA

196 Laux, W. and **Schmidt, G.** 1979. Russische Namen von Arthropoden pflanzenschutzlicher Bedeutung. *Mitteilungen aus der Biologischen Bundesanstalt für Land- und Forstwirtschaft Berlin–Dahlem* **188**: 1–86.

An alphabetical list of Russian names of insects and mites collected from 15 years' work in Russian journals, etc. on plant diseases and plant protection. The Latin equivalents are given with the family name.

SPAIN

197 Bonnemaison, L. 1965. *Enemigos animales de las plantas cultivadas y forestales.* 3 volumes. Ediciones de Occidente; Barcelona.

This is a translation by F. Guerroero of Bonnemaison's original French work, and covers plant and forest pests, which are indexed both under scientific and Spanish common names.

198 Canizo Gomez, J. de and **Arroyo Varela, M.** 1964. Nombres vulgares español de los insectos perjudiciales a las plantas cultivadas. *Boletín de Patología Vegetal y Entomología Agrícola, Madrid* **27**: 101–82.

An excellent and most useful list, indexed under both common and scientific name.

SWITZERLAND

199 Höhn-Ochsner, W. 1976. Zürcher Volkstierkunde. Mundartliche Tiernamen und Volkskundliche Mitteilungen über die Tierwelt des Kantons Zürich. *Vierteljahrsschrift der Naturforschenden Gesellschaft in Zürich* **121**: 15–40 [Insects].

The publication covers all groups of animals, but the above pages are devoted to insects only. The common names are followed by Latin names. It also includes local variations of names from different Swiss cantons.

UK

200 Seymour, P. 1980. *Invertebrates of economic importance in Britain: common and scientific names.* viii + 132pp. HMSO; London.

This publication supersedes and updates earlier lists of names published by government departments in the UK. Although the volume deals also with annelids, nematodes, platyhelminths and molluscs, the larger proportion deals with insects. Entries are restricted to British pests of plants and domestic animals, stored products and timber. The entries are indexed under both common and scientific names.

USA

201 Pfad, R.E. 1978. *Fundamentals of applied entomology.* 798pp. Macmillan; London.

Pages 725–46 of this book are an appendix of common and scientific names. They are listed under common names, with their Latin equivalents, but not cross-referenced.

202 Sutherland, D.W.S. (Ed.). 1978. *Common names of insects and related organisms.* 132pp. Entomological Society of America; College Park, Maryland.

The work has three main sections: common names, scientific names and higher taxonomic categories. Insect and other invertebrate names are included mostly on the basis of their abundance and economic importance. This separately published work replaces papers issued periodically in the *Bulletin of the Entomological Society of America*. It is updated at irregular intervals in that journal.

URUGUAY

203 Trujillo Peluffo, A. 1942. *Insectos y otros parásitos de la agricultura y sus productos en el Uruguay.* 323pp. Alfa; Montevideo.

Indexed by scientific and common names.

USSR – *see* Estonia, Russia

VENEZUELA

204 Guagliumi, P. 1967. Insetti e aracnidi delle piante comuni de Venezuela segnalati nel periodo 1938–1963. *Relazioni e Monografie Agrarie Subtropicali e Tropicali* No. **86** (1966). xix + 391pp.

Largely a list of plants, timbers and stored products showing the insects and mites infesting them in Venezuela within the period covered. This is followed by a list of beneficial insects and arachnids, including their forms of activity. There are three appendices, one of which gives the Venezuelan common names of the arthropods listed. There is a generic and specific index to the Latin names, and a list of authors' names showing their correct abbreviations.

DICTIONARIES AND GLOSSARIES OF ENTOMOLOGY

Glossaries for the subject of entomology are many and varied; entomology is well served in this area of reference material. Dictionaries and glossaries are available in several languages. Frequently, short glossaries are appended to papers on a specific subject, or added to a textbook; but the most commonly used and requested are those of a much more general nature, and it is these, on the whole, that are listed here.

205 Brown, R.W. 1954. *Composition of scientific words.* 882pp. Privately published. Reprinted by the Smithsonian Institution, 1979.

Gives Latin, Greek and other derivations of commonly used scientific words. It is well cross-referenced between classical words, prefixes and suffixes and the English vernacular under which the entries are primarily arranged. The work is of particular use to taxonomists for whom it is a superb source from which to coin Latin or latinized names for new taxa. It is useful also for determining the gender of scientific names, for example when this is required to satisfy the mandatory article of the International Code of Zoological Nomenclature that requires agreement in gender between specific adjectives and generic names.

206 Carvalho, M.B. and **Arrunda, G.P. de.** 1967. Glossário de temos técnicos de entomologia. *Boletim Técnico Instituto de Pesquisas Agronomicas de Pernambuco* **24**: 1–87.

A rather poor list. The pagination makes it appear long and comprehensive, but it is extravagantly laid out. However, there is nothing else serving the Portuguese language.

207 Ericson, R.O. 1961. A glossary of some foreign-language terms in entomology. *Agriculture Handbook, United States Department of Agriculture* No. **218**: 1–59.

Each foreign term indexed is followed by a symbol indicating the language concerned. Russian is transliterated. Although this is a relatively short list of terms, many of the more common words are listed. A few taxonomic characters are diagrammatically illustrated at the end of the work, giving the different language equivalents.

208 Foote, R.H. 1977. *Thesaurus of entomology.* 188pp. Entomological Society of America; College Park, Maryland.

Not a glossary in the strict definition, but it is an excellent list for standardizing terminology in entomology.

209 Ghidini, G.M. *Glossario di entomologia.* 1206pp. La Scuola; Brescia.

A moderately useful dictionary of Italian entomological terms.

210 Hanson, H. 1962. *Dictionary of ecology.* 382pp. Peter Owen; London.

A useful list of terms, but in need of updating.

211 Harbach, R.E. and **Knight, K.L.** 1980. *Taxonomists' glossary of mosquito anatomy.* xi + 415pp. Plexus; Marlton, New Jersey.

A comprehensive treatment of the nomenclature applied to the sclerotized anatomy of the mosquito. Reference is also given to figures for all accepted terms. Well illustrated, and a useful work for student and researcher alike. *See* entry 212.

212 Harbach, R.E. and **Knight, K.L.** 1982. Corrections and additions [to above]. *Mosquito Systematics* **13**: 201–17

213 Harris, R.A. A glossary of surface sculpturing. *Occasional papers in Entomology* No. **28**: 1–31. Department of Food and Agriculture; California.

Forty-four photographs illustrating terms used for surface sculpturing in insects.

214 Jacobs, W. 1975. *Wörterbücher der Biologie. Systematische Zoologie: Insekten.* 377pp. Gustav Fischer Verlag; Stuttgart.

A most useful and well illustrated dictionary. One of a series which cover the field of biology. Each volume is retailed separately.

215 Jaeger, E.C. 1978. *A source book of biological names and terms.* 2nd edn. 323pp. C.C. Thomas; Springfield, Illinois.

A new edition of this most useful book. This edition has a supplement of more than a thousand entries over the older edition. Its content and style are similar to that of Brown (1954), but this book is perhaps not quite so comprehensive in its coverage. It has a short appendix giving brief biographies of men in whose honour commemorative names have been given.

216 Jeffrey, C. 1973. *Biological nomenclature.* 69pp. Systematics Association; London.

Pages 53–69 are a glossary of terms used in nomenclature.

217 Kéler, S. von. 1963. *Entomologisches Wörterbuch mit besonderer Berücksichtigung der morphologischen Terminologie.* 774pp. Akademie Verlag; Berlin.

An excellent, well illustrated entomological dictionary for the German language.

218 King, R.C. 1974. *A dictionary of genetics.* 2nd edn. 375pp. Oxford University Press; Oxford.

A good basic dictionary for the study of genetics. Three useful appendices are also added: appendix A is a chronology showing the order in which certain events having a bearing on genetics took place; appendix B is a list of the periodicals that have often contained articles orientated towards genetics and cytology, and appendix D is a guide to teaching aids.

219 Laliberte, J.-L. 1982. Aide-mémoire à l'usage de l'amateur. Glossaire entomologique. *Fabreries* (Suppl.) **11**: 1–84.

Rather poorly produced in a little-known journal, but a most useful glossary, which deserves to have a good circulation. The glossary is aimed solely at the amateur; it has a good range of terms with simple, basic explanations.

220 Leftwich, A.W. 1976. *A dictionary of entomology.* 359pp. Constable; London.

A poor dictionary, but one which has appeared on many shelves. It should be used with great caution as there are many errors in the definitions given.

221 Lewis, W.H. 1977. *Ecology field glossary: a naturalists' vocabulary.* 152pp. Greenwood Press; Westport, Connecticut.

A small but useful glossary, likely to be useful to the field entomologist. The work also includes an appendix covering measurements, equivalents and conversions.

222 Mayr, E. 1969. *Principles of systematic zoology.* x + 428pp. McGraw-Hill; London.

A glossary useful in the field of systematics can be found on pp. 397–414.

223 Pume, N.D. and **Magnicki, A.W.** 1970. *Ośmiojezyczny stownik rolniczy.* [Agricultural dictionary in eight languages]. 2 volumes. Państwowe Wydawnictwo Rolnicze; Warsaw.

Volume 1 is a single numbered listing of agricultural terms and their equivalents, in Russian, Bulgarian, Czech, Polish, Hungarian, Romanian, German and English. Volume 2 includes a separate alphabetical index for each language referring the reader to the appropriate entry in vol. 1. Includes some entomology and a few common names.

224 Ricker, W.E. 1973. Russian–English dictionary for students of fisheries and aquatic biology. *Bulletin of the Fisheries Research Board, Canada* **183**: 1–428.

A most useful work for translating specialized terms from Russian to English for aquatic insects. Although written primarily for fisheries students, many entomological terms are included, including some common names for which it gives the Latin equivalents. The 'index' is in fact an index to English names, both common and scientific, giving page references where the name can be found with its Russian equivalent.

225   Rieger, R., Michaelis, A. and **Green, M.M.** 1976. *Glossary of genetics and cytogenetics: classical and molecular.* 4th edn. 647pp. Springer-Verlag; Berlin.

Many entomologists are now having to concern themselves with studies of and with genetics. This is a useful glossary with material suitable for students and research workers. The work has also been translated into Russian and Polish.

226   Saether, O.A. 1980. Glossary of Chironomid morphology terminology (Diptera: Chronomidae). *Entomologica Scandinavica (Suppl.)* **14**: 1–51.

Glossary divided into parts for imagines, pupae and larvae. Well illustrated and contains a good bibliography.

227   Séguy, E. 1967. *Dictionnaire des termes d'entomologie.* 465pp. P. Lechevalier; Paris. (*Encyclopédie entomologique* No. 41.)

The standard dictionary for French entomological terms. It is good and reliable.

228   Shchegholeva, V.N. (Ed.). 1955. *Entomologist's handbook and dictionary.* [In Russian]. 451pp. State Publishing House; Moscow.

A moderately useful dictionary for the Russian language. No illustrations.

229   Shiraki, T. 1971. *A glossary of entomology.* 1098pp. Hokuryu-Kan; Tokyo.

Part 1 is a good glossary of English entomological terms and their Japanese equivalents. Part 2 is an alphabetical list of common and specific entomological names in several languages, including English, French and German with Japanese equivalents. Part 3 is a Japanese index in the two Japanese alphabets: Hanji and Kanji.

230   Steinmann, H. and **Zombori, L.** 1981. *Terminologia insectorum morphologica.* 210pp. (*Fauna Hungariae* No. 146.)

A glossary of morphological and anatomical terms. The work is arranged in one alphabetical sequence of Hungarian and scientific names with their relevant equivalents. Seventeen pages of illustrations are included.

231   Torre-Bueno, J.R. de la. 1962. *A glossary of entomology.* 336pp.

Supplement A by G.S. Tulloch. 36pp. Brooklyn Entomological Society; Brooklyn, New York.

An excellent glossary for English-language terms, probably the best of its kind.

232   Torre-Bueno, J.R. de la. 1965. *A glossary of entomology.* 385pp. Science Press; Peking.

A translation into Chinese of entry 231. Pages 293–383 are a transliterated Chinese index.

233   Tuxen, S.L. (Ed.). 1970. *Taxonomist's glossary of genitalia in insects.* 2nd edn. 359pp. Munksgaard; Copenhagen.

There is a chapter on each of the insect orders discussing their genitalia. Pages 217–359 contain the glossary proper. The work includes a good bibliography.

234   Vanekova, Z. (Ed.). 1968. *Skodlive cinitele v pol'nohospodárskej a lesnej výrobe.* [Harmful agents in agriculture and forestry]. 140pp. Chemické Závody Juraja Dimitrova; Bratislava.

Lists the scientific and common names, in Russian, Slovak, Czech, German and English of the 'most important animal pests in agriculture and forestry'. The main listing is in alphabetical order of scientific names, with each entry numbered for ease of reference from the individual indexes which are provided in each language. Mainly entomological, fairly comprehensible and easy to use.

3 Specimens and collections

LABORATORY CULTURES AND MASS REARING OF INSECTS

The mass rearing of insects is an important part of the experimental work done with insects, particularly in the field of biological control, and other experimental work with economic pests. Some of the books published around this subject area become quickly out of date – but they are often good starting points, and most contain good basic information.

235 **Dickerson, W.A., Hoffman, J.D., King, E.G., Leppla, N.C.** and **Odell, T.M.** *Arthropod species in culture in the United States and other countries.* 93pp. Entomological Society of America; College Park, Maryland.

In tabular form. The location of species collections is given, as well as the diet and availability.

236 **Galtsoff, P.S., Lutz, F.E., Welch, P.S.** and **Needham, J.G.** 1973. *Culture methods for invertebrate animals.* xxxii + 590pp. Comstock; Ithaca, New York.

Although now rather dated, this perhaps is the most comprehensive guide of its kind, covering more than 60 species of insects.

237 **Singh, P.** **1977.** *Artificial diets for insects, mites and spiders.* 594pp. Plenum; New York and London.

The work is arranged by order of insect. The author quotes the paper in which the diet was set in a 'rearing' paper and then describes in some detail the diet used. A most useful work, forming both a bibliography and a compendium of diets for rearing of insects.

238 **Singh, P.** and **Charles, J.G.** 1975. A list of laboratory cultures and rearing methods of terrestrial arthropods in New Zealand. *Bulletin of the Entomological Society of New Zealand* **3**: 1–30.

The work describes simple methods for rearing laboratory species, mainly for experimental work with economic pests.

239   Skuhravy, V. 1968. *Metody chovu hymyzu* [Methods of insect breeding]. [In Czech]. 285pp. Academia, Československá Akademie Věd; Prague.

The work describes in some detail methods for breeding insects in the laboratory. The text is arranged under families of insects.

240   Smith, C.N. (Ed.). 1966. *Insect colonization and mass production.* 618pp. Academic Press; New York.

The work describes the production of insect colonies for major research on insect control work. Multiple authorship of the work has ensured coverage of all the major groups of animals.

241   UFAW (Ed.). 1972. *The UFAW handbook on the care and management of laboratory animals.* 4th edn. x + 624pp. Churchill Livingstone; Edinburgh and London.

Nine of the 53 chapters in this authoritative work are concerned with rearing insects for laboratory use, either for study in themselves or as food for other animals. An excellent work, but should ideally be read in conjunction with the 2nd edition (entry 243), to which frequent reference is made.

242   Wong, H.R. 1972. *Literature guide to methods for rearing insects and mites.* Northern Forest Research Centre; Edmonton, Alberta. (Information Report NOR-X-38.)

A very full bibliography on the subject – now somewhat out of date, but still a useful starting point. Indexed by order of insect and then alphabetically by author.

243   Worden, A.N. and **Lane-Petter, W.** (Eds). 1957. *The UFAW handbook on the care and management of laboratory animals.* 2nd edn. xix + 951pp. AFAW; London.

Although superseded by the 3rd and 4th editions, this volume still contains much useful information, to which frequent references are made in the later editions.

244   Wyniger, R. 1974. *Insektenzucht: Methoden der Zucht und Haltung von Insekten und Milben im Laboratorium.* 368pp. Verlag Eugen Ulmer; Stuttgart.

Laboratory techniques of collecting and rearing insects and mites are described in some detail.

COLLECTION, CURATION AND PRESERVATION

The conspicuous decline of butterfly populations, especially in Britain, and the vulnerability of these insects to collectors has led to conservation movements aimed at the preservation of particular species or special insect habitats. Collecting is less fashionable than it was, but is still essential for scientific purposes where certain groups or regional insect faunas remain understudied. Nowadays there is even a code of practice (entry 255).

Several publications provide guidance both for the professional and the amateur entomologist on the collection and rearing of insects and the best methods for their preservation and study. Such works can be specially helpful to those participating in collecting expeditions who cannot be familiar with the preferred methods appropriate to every insect group. The listed works cover insects in general. Detailed guidance for particular insects can often be found in the specialist literature, for example medically important insects in Smith's book *Insects and other arthropods of medical importance* (entry 810).

245   Cogan, B.H. and **Smith, K.G.V.** *Insects: instructions for collectors.* No. 4a. 169pp. British Museum (Natural History); London.

Advice and instructions on preferred methods for collecting and preserving specimens for museum collections. It is particularly useful for the expedition leader or person responsible for returning specimens from similar trips in good condition. It is also a useful guide for the beginner.

246   Dahms, E., Monteith, G. and **Monteith, S.** 1979. Collecting, preserving and classifying insects. *Queensland Museum Booklets* No. **13**: 1–28.

Written mainly for secondary-school biology courses, it is also a useful guide for beginners. Poorly illustrated.

247   Ford, R.L.E. 1973. *Studying insects: a practical guide.* 150pp. Frederick Warne; London.

Written mainly for the amateur, having the study of insects as a hobby.

248   Martin, J.E.H. 1977. *Collecting, preparing and preserving insects, mites and spiders.* 182pp. Agriculture Canada; Ottawa. (*The insects and arachnids of Canada*, Pt. 1.)

A most useful volume, covering all the arthropod groups. The work is clearly illustrated, showing a number of accepted methods for the collection and preparation of insects in particular.

249   Norris, K. R. and **Upton, M.S.** 1974. The collection and preservation of insects. *Miscellaneous Publications of the Australian Entomological Society* No. **3**: 1–33.

Brief descriptions of methods of collecting and preserving insects. Descriptions are not so detailed as in entry 248 or 250, for example.

250   Oldroyd, H. 1970. *Collecting, preserving and studying insects.* 336pp. Hutchinson; London.

A good text for both the amateur, student and professional. The work is rather poorly illustrated, except for the photographs. The volume is well indexed and has a good bibliography.

251   Peterson, A. 1959. *Entomological techniques: how to work with insects.* 435pp. Edwards; Ann Arbor, Michigan.

Aimed at both the professional and the amateur. The work has many good line drawings, and is a very detailed study of the subject.

252   Smithers, C. 1982. *Handbook of insect collecting: collection, preparation, preservation and storage.* 120pp. David & Charles; Newton Abbot, England.

An introductory guide for making insect collections, including a key to identify the orders of insects. There is a most useful synopsis of killing and preserving techniques for each order. Generally the illustrations are poor.

253   Wagstaff, R. and **Fidler, J. H.** (Eds). 1970. *The preservation of natural history specimens.* 3 volumes. Vol. 1, *Invertebrates.* 205pp. H.F. & G. Witherby; London.

Volume 1 of this three-volume work deals almost exclusively with insects. It lacks some of the more up-to-date styles of mounting some of the insect groups, but it is a comprehensive study for most groups of insects.

254   Walker, A. K. and **Crosby, T. K.** 1979. The preparation and curation of insects. *DSIR Information Series* **130**: 1–55.

Explains thoroughly the methods of preparing insects for collections and also how the collection should be curated and managed.

COLLECTING CODES

255   *A code for insect collecting.* Joint Committee for Conservation of British Insects, c/o Royal Entomological Society, 41 Queen's Gate, London SW7 5HU, England.

256   *Nature photographers' code of practice.* Association of Natural History Photographic Societies, Royal Society for the Protection of Birds, The Lodge, Sandy, Bedfordshire, England.

257   Dunn, G.A. 1982. A directory of policies on arthropod collecting on public lands. *Great Lakes Entomologist* **15**: 123–41.

This directory serves as a guide to the policies and regulations on arthropod collecting and research on public lands in the USA. The list is arranged by US states alphabetically. Addresses are given from which written permission is required, and the procedures to be adopted are explained.

258   McGaugh, M.H. and **Genoways, H.H.** 1976. State laws as they pertain to scientific collecting pursuits. *Museology* **2**: 1–81.

Lists requirements and regulations of each state in the USA. The protected and rare and endangered species for each state are also listed.

LOCATION OF COLLECTIONS

Over the years the collections of many naturalists and entomologists have been dispersed throughout the world. Famous collections rich in type-material and often needed by the specialist for inspection or for the borrowing of type-material for comparison, can be difficult to trace. Many of these collections have been bought and sold, often more than once; sometimes they have been given away to other individuals, who incorporated them into their own collections. Fortunately for the researcher and specialist most of these collections – at least, the more important of them – have eventually found themselves in the larger museums and research institutes of the world. Some important works have been compiled for tracing these collections.

259 Chalmers-Hunt, J.M. 1976. *Natural history auctions 1700–1972: register of sales in the British Isles.* xii + 189pp. Sotheby Parke Bernet; London.

An excellent reference work for information on the sale of collections, and where sale catalogues are deposited. Often regarded as ephemera, sale catalogues can be a most valuable source for tracing collections. The sales are listed chronologically, perhaps not the easiest style for the user, but the work is well indexed.

260 Horn, W. and **Kahle, I.** 1937. Über entomologische Sammlungen Entomologen und Entomo-Museologie. *Entomologische Beihefte* **2–5**: vi + 536pp., 38 pls.

The work is arranged under the name of the collector, and gives details of each collector's full name and lifespan dates, along with the whereabouts of the collection and the group(s) of insects it contains. The plates of this work contain examples of labels from specimens in the handwriting of well-known entomologists, naturalists and collectors. Pages 313–88 are a supplement with addenda and corrigenda. The work is well indexed. There is a supplement by Sachtleben (entry 261).

261 Sachtleben, H. 1961. Nachtäge zu Walter Horn and Ilse Kahle: 'Über entomologische Sammlungen'. *Beiträge zur Entomologie* **11**: 481–540.

As its title suggests, this is a supplement and continuation of the original work and arranged in exactly the same way.

262 Sherborn, C.D. 1940. *Where is the . . . collection? An account of the various natural history collections that have come under the notice of the compiler 1880–1939.* 148pp. Cambridge University Press; Cambridge, England.

A rather more abridged work than that of Horn and Kahle (entry 260), but contains collections other than entomological collections. The list gives the name of the collector and location of his collection. Parts of this work are in the process of being updated for publication.

263 Zimsen, E. 1964. *The type material of I.C. Fabricius.* 656pp. Munksgaard; Copenhagen.

Fabrician types, often searched for, are all listed in this work. Original references are included and where the types are deposited. The work is arranged by insect family, each having its own index.

ENTOMOLOGICAL SUPPLIERS

Although entomological suppliers are not, strictly speaking, an information source, we feel that the inclusion of this list is justified by the number of requests we receive for assistance in obtaining books, equipment, livestock and specimens.

No comprehensive list has, as far as we know, been published previously, and we ourselves have included only a small selection of the many professional and semiprofessional suppliers. Many of the society and group newsletters regularly feature advertisement sections, and these are a rich source of information on further suppliers, particularly for livestock and specimens. Within the UK, the popular weekly magazine *Exchange & Mart* has become the accepted place for the advertisement of insect specimens and livestock. The great majority of the UK suppliers are usually in attendance at the Annual Exhibition of the Amateur Entomologists' Society, usually held near London in the autumn, and the Midlands Entomological Fair and Exhibition, usually held in Leicester during spring. Both of these events are well advertised.

We have not included suppliers of biological control agents, on the grounds that these fall more properly into the area of a similar guide to information sources on pest control. However, useful lists are published from time to time in the *IPM Practitioner* (entry 1126) and the Entomological Society of America is, at the time of writing, collecting a list for possible publication.

It should go almost without saying that the inclusion or exclusion here of a supplier is in no way a recommendation or otherwise. Our aim here, as throughout this guide, is to provide our readers with some useful suggestions.

Key: **B** = Books   **E** = Equipment   **L** = Livestock   **S** = Specimens

264   Arthropod Specialities Co.   **E**
PO Box 1973, Sacramento, California 95809, USA.

Mainly polythene storage vials and other small items.

265   American Biological Supply Co.   **B,E**
1330 Dillon Heights Avenue, Baltimore, Maryland 21228, USA.

266   Australian Entomological Press   **E**
14 Chisholm Street, Greenwich, New South Wales 2065, Australia.

Data labels printed.

267   Australian Entomological Supplies   **B,E**
35 Kiwong Street, Yowie Bay, New South Wales 2228, Australia.

268   R.N. Baxter   **L,S**
45 Chudleigh Crescent, Ilford, Essex IG3 9AT, England.

Mainly Saturniidae and other exotic insects. Mail order only.

269   BioQuip Products   **E**
PO Box 61, Santa Monica, California 90406, USA.

Mainly professional equipment.

270 Butterfly Centre L,S
Plummer, Tenterden, Kent, England.
Mainly tropical Lepidoptera. Visitors by appointment.

271 Butterfly Farm Ltd E,L,S
Bilsington, Near Ashford, Kent TN25 7JW, England.
Mainly Lepidoptera, but some other exotic insects.

272 R. Cattle Ltd E
PO Box 1, Oxford Road, Gerrards Cross, Buckinghamshire, England.
Precision woodworkers, making cabinets and store/display cases.

273 Clo Wind E,S
827 Congress, Pacific Grove, California 93950, USA.

274 L. Christie B,E,S
129 Franciscan Road, Tooting, London SW17 8DS, England.
Mail order only.

275 E.W. Classey Ltd B
PO Box 93, Faringdon, Oxfordshire SN7 7DR, England.
Specialist entomological publisher and bookseller. Particularly good for secondhand books.

276 Combined Scientific S
PO Box 1446, Fort Davis, Texas 79734, USA.
Particularly good for the less 'popular' orders, e.g. Neuroptera.

277 Deyrolle B,E,S
46 rue du Bac, F-75007 Paris, France.

278 Entomological Reprint Specialists B
PO Box 77224, Dockweiler Station, Los Angeles, California 90007, USA.

279 Flatters & Garnett Ltd E
Mikrops House, Bradnor Road, Manchester 22, England.
Good for microscopes.

280 P.D.J. Hugo E
38 Cotswold Crescent, Chipping Norton, Oxfordshire OX7 5DT, England.
Data labels printed.

281 Insect Farming & Trading Agency S
Division of Wildlife, PO Box 129, Bulolo, MP, Papua New Guinea.
A non-profit-making government agency set up to organize and control the export of PNG insects. Mainly butterflies, including the giant birdwings (*Ornithoptera*), many of which are captive-bred.

282 International Butterflies   **S**
4 Alexandra Road, Minehead, Somerset, England.

Mainly Lepidoptera and Coleoptera. Mail order only.

283 R.B. Janson   **B**
44 Great Russell Street, London WC1, England.

New and secondhand books. Specializes in finding customers' wants.

284 K.C. Liew   **S**
Scientific Entomological Emporium, PO Box 56, Taiping, Perak, Malaysia.

285 London Butterfly House   **B,E,L,S**
Syon Park, Brentford, Middlesex, England.

No postal business. The main attraction is a year-round display of free-flying tropical butterflies in a huge walk-through glasshouse.

286 Macmillan Science Co. Inc.   **E**
8200 South Hoyne Avenue, Chicago, Illinois 60620, USA.

287 National Butterfly Museum (formerly Saruman Butterflies)   **B,E,L,S**
St Mary's, Bramber, Sussex, England.

A rather controversially named commercial organization. Exhibition area.

288 Nature-All   **E,S**
David W. Boulton, Box 1181, Wilkes-Barre, Pennsylvania 18703, USA.

Specializes in tropical American Lepidoptera.

289 Nature and Field Hobby Centre   **E**
24 Burlington Street, Crows Nest, New South Wales 2065, Australia.

290 Nature of the World   **B,E,L,S**
Fetcham Cottage, 19 Bell Lane, Fetcham, Leatherhead, Surrey, England.

Exhibition area.

291 Luis E. Pena G.   **S**
PO Box 2974, Santiago 1, Chile.

Specializes in South American insects. Collecting trips undertaken.

292 Q Naturalist Enterprises   **L,S**
69 Marksbury Avenue, Kew, Richmond, Surrey, England.

Mail order only.

293 Queensland Butterfly Company   **L,S**
Long Road, North Tamborine, Queensland 4272, Australia.

Breeders of Australian butterflies and importers of exotic species.

294 Science Aids Pty Ltd E
12 Dequetteville Terrace, Kenttown, South Australia 5067.

295 A. Shutt E,L,S
30 Glover Road, Sheffield, England.
Specimens include some of the less 'popular' orders, e.g. Odonata. Setting boards made to customers' specifications.

296 Tradewing S
82 Balaclava Road, Cairns, North Queensland 4870, Australia.

297 Transworld Butterfly Company B,E,L,S
Apartado 7911, San Jose, Costa Rica, Central America.
Livestock mainly of exotic butterflies (and bird-eating spiders).

298 Watkins & Doncaster B,E,S
Four Throws, Hawkhurst, Kent, England.
Mainly equipment and specimens. Good for cabinets, pins, etc. Showroom.

299 Wheldon & Wesley Ltd B
Lytton Lodge, Codicote, Hitchin, Hertfordshire SG4 8TE, England.
Good for secondhand books.

300 Worldwide Butterflies Ltd B,E,L,S
Compton House, Sherborne, Dorset, England.
Specimens and livestock of a very wide range of British and exotic species of Lepidoptera and other insects. Also houses the Lullingstone Silk Farm, a working *Bombyx mori* farm. Large display area of live butterflies and showroom.

ENTOMOLOGICAL PHOTOGRAPHY

The photographing of insects is becoming an increasingly popular pastime. It may be that the move to conservation rather than collecting, thus to watch rather than collect, has added to its popularity; although it must be noted that most of the books on insect photography do give advice on collecting, rather than on photography in the field.

301 Angel, H. 1975. *Photographing nature – insects.* 96pp. Fountain Press; Kings Langley, England.

The groups of insects are briefly discussed, with advice on how to photograph them. There are many photographic examples. A short appendix includes a checklist of equipment for insect photography and on pp. 92–3 is a brief glossary of photographic terms.

302 Lindsley, P.E. *Insect photography for the amateur.* viii + 52pp. Amateur Entomologists' Society; London.

Describes the necessary equipment and accessories for macro, micro and general insect photography. A good practical guide to the subject.

303   Linssen, E.F. 1953. *Entomological photography in practice.* 112pp. Fountain Press; London.

Discusses the apparatus and equipment needed for insect photography. Some emphasis is laid on photographing immature stages and the problems that may be encountered.

304   Malies, H.M. 1959. *Applied microscopy and photomicrography.* 143pp. Fountain Press; London.

The techniques and equipment for this kind of work are improving all the time, but there is some useful basic information in this work.

305 *Nature photographers' code of practice*

Produced by the Association of Natural History Photographic Societies: RSPB, The Lodge, Sandy, Bedfordshire SG19 2DL, England.

SOURCES FOR ILLUSTRATIVE MATERIAL

We are frequently asked to suggest possible sources or suppliers of photographs, slides or transparencies of various entomological subjects, either purely for study or for publication. Quite apart from the difficulty of tracking down any sort of illustration of some insects, it is often important that the picture should show some special feature or piece of behaviour that can really only be captured using fairly sophisticated photographic techniques of the kind employed by the major professional photographers.

We have assembled in this section a fairly brief list of possible sources, and tried to give an indication of their particular specializations, if any. It should be remembered that in many cases a charge will be made, even for the loan of material, and that this can be quite considerable in the case of the commercial agencies. We must also emphasize that sources included in this list are not necessarily geared towards dealing with large numbers of requests, and that their inclusion here does not necessarily imply any recommendation on our part. However, we hope that our list, used with caution, may be of some help in locating difficult-to-find material, and perhaps also take some of the pressure off the better-known and therefore overworked sources.

Sources for photographs and/or transparencies

306   Ardea London Ltd. 35 Brodrick Road, London SW17 7DX, England.

Specialist natural history photographic agency with good coverage of European insects and those of Australasia. Holds mainly 35 mm colour transparencies.

307   Biofotos. Ms Heather Angel, 'Highways', Vicarage Hill, Farnham, Surrey GU9 8HJ, England.

One of the foremost photographers of entomological subjects in the UK. A range of photographs and 35 mm transparencies are available.

308   British Museum (Natural History)
      Department of Entomology, Orthoptera Section.

A large collection of 35 mm transparencies of live Orthoptera (mostly European, although some tropical species are included). The collection is intended as a record of the natural colouring of these insects for study in the Museum. Loan material may be made available under special circumstances such as for lectures to schools or scientific societies. Application to use the collection should be made to the Head of the Orthoptera Section, Department of Entomology, British Museum (Natural History), Cromwell Road, London SW7 5BD, England.

Mr D. J. Carter, Department of Entomology

Colour transparencies (35 mm) of Lepidoptera; mainly of larvae or other immature stages, and mainly British species.

309 Ciba-Geigy Agrochemicals Ltd. Whittlesford, Cambridge CB2 4QT England.

Mainly 35 mm colour transparencies of veterinary subjects.

310 Centre for Overseas Pest Research. College House, Wrights Lane, London W8 5SJ, England.

Mainly 35 mm colour transparencies of agriculturally important insects, including some relating to mosquito and tsetse control. Particularly strong on Orthoptera, Isoptera, some Hemiptera, some Lepidoptera (especially *Spodoptera*). Material available only on loan.

311 Frank W. Lane. Drummoyne, Southill Lane, Pinner, Middlesex HA5 2EQ, England.

Colour transparencies (35 mm) and black-and-white prints of bees, butterflies and moths.

312 Glasshouse Crops Research Institute. Worthing Road, Rustington, Littlehampton, West Sussex BN16 3PU, England.

Colour transparencies and black-and-white prints of the pests of glasshouse crops and their natural enemies.

313 John Topham Picture Library. Edells Markbeech, Edenbridge, Kent TN8 5PB, England.

Colour transparencies (35 mm) and black-and-white prints of British insects.

314 London Scientific Fotos Ltd. 109–110 Bolsover Street, London W1P 7HF, England.

Specialize in high-quality photomicrographs and photomacrographs of invertebrates, including behavioural sequences. Both agricultural and medical/veterinary subjects are held, mainly as 35 mm colour transparencies. A catalogue is available.

315 Ministry of Agriculture, Fisheries and Food. Agricultural Development Advisory Service (ADAS), Great Westminster House, Horseferry Road, London SW1P 2AE, England.

A very comprehensive collection including many entomological subjects. These have been selected from those held by MAFF entomologists, so there may be some duplication of material available from individual MAFF laboratories. A well-indexed catalogue is available.

316   Ministry of Agriculture, Fisheries and Food, Harpenden Laboratory. Hatching Green, Harpenden, Hertfordshire AL5 2BD, England.

Colour transparencies (35 mm) and black-and-white photographs of agriculturally important pests.

317   Ministry of Agriculture, Fisheries and Food, Slough Laboratory. London Road, Slough SL3 7HJ, England.

Black-and-white photographs (including scanning electron micrographs) of storage pests.

318   Murphy Chemicals Ltd. Wheathampstead, St Albans, Hertfordshire AL4 8QU, England.

Colour transparencies (35 mm) of pests of agricultural and garden crops and plants, although emphasis is primarily on the damage caused rather than the pests.

319   National Vegetable Research Station. Wellesbourne, Warwick CV35 9EF, England.

Colour transparencies (35 mm) and black-and-white prints of pests of vegetables.

320   Natural History Photographic Agency. Moray Lodge, Sandling Road, Saltwood, Near Hythe, Kent, England.

Run by L. Hugh Newmann, this professional agency specializes in tracing illustrative material to order rather than maintaining a reference collection. Acts as agency for many of the very best insect photographers.

321   Natural Science Photos. 33 Woodland Drive, Watford, Hertfordshire WD1 3BY, England.

Accurately identified, well-documented colour transparencies. All areas of natural science are covered, including insects. Special commissions undertaken.

322   Oxford Scientific Films Ltd. Lower Road, Long Hanborough, Oxford OX7 2LD, England.

Mainly 35 mm colour transparencies of arachnids, forest and desert insects, flight pictures and life histories. A commercial picture library holding some quite unique and dramatic material.

323   Rentokil Ltd. PR Department, Rentokil House, Garland Road, East Grinstead, West Sussex RH19 1DY, England.

Colour transparencies (35 mm) and black-and-white photographs of wood-boring insects, household pests and pests of stored products. Also holds 16 mm cine films on termites, cockroaches and woodworm.

324   Ries Memorial Slide Library. Entomological Society of America, 4603 Calvert Road, College Park, Maryland 20740, USA.

A collection of approximately 1500 colour slides of insects and other arthropods, listed in *Bulletin of the Entomological Society of America* **26**: 452–63 (1980). Slides are available for private or non-profit, educational purposes or for reproduction in publications.

325   Royal Horticultural Society. Entomology Department, RHS Garden, Wisley, Woking, Surrey GU23 6QB, England.

Colour transparencies (35 mm) and black-and-white photographs of horticultural pests and the damage they cause.

326   Science Photo Library. 2 Blenheim Crescent, London W11 1NN, England.

Colour transparencies (35 mm) and black-and-white photographs of common and medically important insects, especially scanning electron micrographs.

327   Shell Photographic Library. The Shell Centre, London SE1 7NA, England.

Transparencies, slides and prints, in colour and black-and-white, of a large number of plant pests. A catalogue is available.

328   Spectrum Colour Library. 146 Oxford Street, London W1, England.

Mainly 35 mm colour transparencies of a wide range of natural history subjects, including butterflies and moths.

329   Tropical Products Institute. Public Relations Section, 56–62 Gray's Inn Road, London WC1X 8LU, England.

Material illustrating pests of tropical products, including those of stored grains, and of the damage they cause.

330   Department of Agriculture. Photograph Library, 14th and Independence Ave. SW, Washington, DC 20250, USA.

Holds a very large collection of black-and-white photographs and colour transparencies, covering all aspects of agriculture including applied entomology.

331   Van Emden, H.F. c/o Department of Agriculture and Horticulture, The University, Reading RG6 2AH, England.

A private collection of slides showing agriculturally important pests and their parasites and predators. Particularly strong on pests of grain legumes.

332   Worldwide Butterflies Ltd. Compton House, Sherborne, Dorset DT9 4QN, England.

Colour plates (not transparencies) of Lepidoptera, including many exotic species but not generally those of economic importance.

Other sources of illustrative material

The following textbooks contain unusual or exceptionally good photographs or plates.

333 **[Anon.]** 1979. *Insects: an illustrated survey of the most successful animals on earth.* 240pp. Hamlyn; London.

Six internationally known entomologists contributed to this general work on entomology. The illustrations are very good, and there is a large selection of particularly fine colour photographs showing all stages of insects from eggs to adults.

334 **Bandsma, A.T.** and **Brandt, R.T.** 1963. *The amazing world of insects.* 46pp. 133 plates. George Allen & Unwin; London.

All black-and-white photographs, mainly taken at unusual angles.

335 **Bohlen, E.** 1978. *Crop pests in Tanzania and their control.* 2nd edn. 142pp. 252 colour photographs. Paul Parey; Berlin and Hamburg.

Very good colour illustrations of Tanzanian crop pests.

336 **Conci, C.** and **Bocciarelli, I.** 1968. *Meraviglie della natura. Insetti.* 155pp. Istituto Geografico de Agostini; Novara.

Many fine colour and black-and-white photographs of all groups of insects.

> **Dalton, S.** *Borne on the wind: the extraordinary world of insects in flight* – see entry 697

337 **Engel, H.** (Ed.). 1961. *Sammlung Naturkundlicher Tafeln: Mitteleuropäische Insekten.* 192 plates. Kronen-Verlag Erich Cramer; Hamburg.

A collection of 192 coloured plates of insects and some spiders and centipedes. The verso of the plate gives a description of the orders or families represented on the plate (usually more than one). Gives German common names, and some information on life history and distribution.

338 **Hyde, G.E.** 1973–. *Insects in Britain,* books 1–4. *British moths,* books 1–4. *British caterpillars (moths),* books 1–2; *(butterflies),* 1 book. *British butterflies,* books 1–2. Jarrold Colour Publications; Norwich.

Small, popular booklets with good photographs of most of the indigenous species.

339 **Klots, A.B.** and **Klots, E.B.** 1959. *Living insects of the world.* Hamish Hamilton; London.

Heavily illustrated with colour and black-and-white photographs.

340 **Kranz, J., Schmutterer, H.** and **Koch, W.** 1977. *Diseases, pests and weeds in tropical crops.* 666pp. Paul Parey; Berlin and Hamburg.

Includes some particularly good colour photographs of insect pests.

341 Manuel, J. and **Coutin, R.** 1978–. *Atlas des insectes*; Tom 1: *Les coléoptères*; Tom 2, *Les papillons*. Leden; Paris.

All species described are accompanied by full colour photographs.

342 Sandhall, A. 1975. *Insects and other invertebrates in colour.* 204pp. Lutterworth; Guildford and London.

A popular book, mostly on insects and with a good selection of colour plates, showing the subjects in their natural habitats. Also includes a simple glossary.

343 Sakaguti, K. 1980–. *Insects of the world.* Approx. 6 volumes. Hoikusha; Osaka.

Although the text is in Japanese, there are Latin legends to the very fine colour plates. The work is divided geographically: vol. 1 deals with south-east Asia plus Australia, etc.

344 Sandved, K.B. and **Brewer, J.** 1980. *Butterflies.* 176pp. Harry N. Abrams; New York.

A semi-popular book, but the plates include some unusual and good photographs of scales of wings of the different families. A few early stages are also shown, but most of the pictures illustrate adults.

345 Scortecci, G. 1966. *Insetti.* 2 volumes. Edizione Labor; Milan.

A useful text on Italian entomology; the work has numerous colour and black-and-white photographs of insects, usually in their natural habitat.

Skaife, S.H. *African insect life — see* entry 89

346 Stanek, V.J. 1969. *The pictorial encyclopedia of insects.* 544pp. Hamlyn; London.

Includes more than a thousand black-and-white and colour photographs of a very wide cross-section of the insect world. Well indexed.

Further sources

The Association of British Picture Libraries and Agencies (ABPLA, PO Box 93, London NW6 5XW, England) publishes a useful directory of its members, with a brief description of the types of pictures, transparencies and services available from each.

347 Evans, H., Evans, M. and **Nelki, A.** 1975. *The picture researcher's handbook: an international guide to picture sources — and how to use them.* 365pp. David & Charles; Newton Abbot, England.

The volume is divided into three main sections: part 1, a general guide on how to use the volume; part 2, a directory of sources, and part 3, indexes, including a subject and geographical index.

4 The literature of entomology

PRIMARY JOURNALS, REVIEW JOURNALS, MONOGRAPH SERIES

Although it will become apparent from the next section that entomological book publishing is a flourishing concern, the overwhelming bulk of entomological information is published in the form of papers in scientific journals and other periodical literature. This is dramatically illustrated in the case of applied entomology, where an average of more than 90% of the literature abstracted in *Review of Applied Entomology* is of this kind.

One of the major problems facing the newcomer to the serial literature of entomology is the selection and possibly acquisition of relevant titles from the vast number being published. We have estimated that there may be some 1600 or so titles which at least occasionally include papers on some aspect of the subject. From these, we have selected this list of about 250 which we feel to be the most useful.

For each title we have given brief notes including: start date; number of issues per year; languages used; address of supplier or publisher; and a short summary of contents and any special features such as book reviews or advertisements.

Although for completeness we have included a few of the more important defunct titles, we have on the whole restricted our list to those which are currently active. The older literature is so extensive that it is perhaps best approached through the various abstracting and indexing journals and subject bibliographies.

Readers who wish to consult journals listed here should refer to the section 'Locating source documents', p. 162.

348 Acrida. 1972–81. 4 per year. En, Fr. Association d'Acridologie, 105 boulevard Raspail, 75 Paris 6, France.

Original papers on all aspects of acridology. Also includes *Acridological Abstracts*, comprising informative abstracts of world literature, prepared by the Centre for Overseas Pest Research (UK).

349 Acta Entomologica Bohemoslovaca. 1956–. 6 per year. En, Cz, Ru, De. Institute of Entomology, Czechoslovak Academy of Sciences, U Šalamounky 41, 15800 Praha 5, Czechoslovakia.

Taxonomy of the Palaearctic fauna. About 2% applied entomology. Carries book reviews, biographies and bibliographies.

350 Acta Entomologica Fennica (Suomen Hyönteistieteellinen Seura). 1947–. Irregular. En, Fi. Entomological Society of Finland, Pohj. Rautatiekatu 13, SF-00100 Helsinki 10, Finland.

Monographic series.

351 Acta Entomologica Jugoslavica (and supplements). 1920–. (ceased 1931–71). 4 per year. En, Yu, De. Jugoslavensko Entomološko Društvo, pret (PO Box) 360, Yu-41001 Zagreb, Yugoslavia.

Original research on all branches of entomology, particularly on Yugoslavian fauna. Rarely issued on time. Book reviews.

352 Acta Entomologica Lituanica. 1970–. 2 per year. Li, Ru. Academy of Sciences of the Lithuanian SSR, Institute of Zoology and Parasitology, Lithuanian Entomological Society, Vilnius, Lithuanian SSR.

Original research. The content of some volumes is made up of papers contributing to a general theme, thus: Vol. 3, *Tree and shrub pests and their control.*

353 Acta Entomologica Musei Nationalis Pragae (and supplements) **(Sborník Entomolgickeho Odděleni Národního Muzea v Praze).** 1923–. 1 per year. En, Fr,· De, Cz. Národní Muzeum v Praze, Přirorvědecké Muzeum, 14800 Praha 4, Czechoslovakia.

Mostly entomology of the Palaearctic region. Many papers on reports of expeditions. The supplements are usually monographic papers. The issue of this journal has become somewhat irregular.

354 Acta Entomologica Sinica. 1954– (ceased 1967–72). 4 per year. Ch, En. Academy of Sciences, Institute of Zoology, Peking, China.

Original research and brief communications. Includes medical entomology and papers on acarina. A cover-to-cover English translation was issued for one volume only (1968) and then discontinued.

355 Acta Faunistica Entomologica Musei Nationalis Pragae (and supplements). 1956–. Irregular. Cz, En, De. Národní Muzeum v Praze, 14800 Praha 4, Czechoslovakia.

Mongolian–Czechoslovakian expedition reports. Results of other Czech expeditions. Systematics of Czechoslovakian entomological fauna. In recent years, volumes have been published rather irregularly.

356 Acta Oecologica, Oecologia Applicata. 1980–. 4 per year. Fr, En, De, Es. CDR Centrale des Revues, BP 119, F-93104 Montreuil Cedex, France.

Devoted to research papers relating to the application of ecological concepts, particularly in the animal world, and more generally to research leading to the management of man-modified ecosystems. Comprises 80% economic entomology.

357 Acta Phytopathologica Academiae Scientarum Hungaricae. 1966–. 4 per year. En. Academiae Scientarum Hungaricae, PO Box 509, Budapest 24, Hungary.

Original papers on phytopathology including 20% on insect pests; book reviews.

358 Acta Tropica. 1944–. 4 per year. En, De, Fr. Schwabe & Co., Steinentorstr. 13, CH-4010 Basel, Switzerland.

Original papers, review articles and short communications on all aspects of biomedical sciences in the tropics, including 15% medical entomology.

359 Acta Zootaxonomica Sinica. 1964–. 4 per year. Ch, En. Academy of Sciences, Institute of Zoology, Peking, China.

Original research, some 60% on entomology. Taxonomy and faunal studies.

360 Advances in Insect Physiology. 1963–. 1 per year. En. Academic Press, 24–28 Oval Road, London NW1 7DX, England.

Insect physiology interpreted in its broadest terms. Each volume contains comprehensive surveys on selected topics. As with most review articles the bibliographies are excellent.

361 Agricultural Gazette of New South Wales. 1890–. 6 per year. En. Department of Agriculture, McKell Building, Rawson Place, Sydney, New South Wales 2000, Australia.

Brief reviews and recommendations on various agricultural topics, including 12% agricultural/veterinary entomology. Semi-popular style. Annual author/subject index.

362 Agricultural Research Journal of Kerala. 1963–. 2 per year. En. Kerala Agricultural University, Nellanikkara 680654, Keɪala, India.

Original research in all fields of agricultural science and new pest records, including 25% agricultural entomology.

363 Alexanor (Revue des Lépidoptéristes Français). 1959–. 4 per year. Fr. Alexanor, 45 bis rue de Buffon, F-75005 Paris, France.

Systematics, biology and distribution of French Lepidoptera; book reviews, annual index.

364 American Journal of Tropical Medicine and Hygiene. 1952–. 6 per year. En. Allen Press, 1041 New Hampshire Street, Lawrence, Kansas 66044, USA.

Original research, brief communications and correspondence on tropical medicine and hygiene, including 15% medical entomology; book reviews. Annual author/subject indexes.

365   Anais da Sociedade Entomológica do Brasil. 1972–. 2 per year. Pt, Fr, En, Es. Sociedade Entomológica do Brasil, A/C Octavio Nakano, Dpto de Entomologia – Esalq – USP, Caixa Postal 9, 13.400 Piracicaba, SP, Brazil.

All aspects of entomology including articles on economic pests.

366   Anales del Instituto Nacional de Investigaciones Agrarias, Serie Proteccion Vegetal. 1971–. Irregular. Es, En. Instituto Nacional de Investigaciones Agrarias, General Sanjurjo 56, Madrid 3, Spain.

Original research on Spanish plant protection, including 60% agricultural entomology.

367   Angewandte Parasitologie. 1960–. 4 per year. De, Ru, En. VEB Gustav Fischer Verlag, Villengang 2, 69 Jena, German Democratic Republic.

Original research on applied parasitology including 10% economic entomology.

368   Annales Entomologici Fennici (Suomen Hyönteistieteellinen Aikakauskirja). 1935–. 10 per year. Fi, En. Department of Agricultural and Forest Zoology, University of Helsinki, SF-00710 Helsinki 71, Finland.

General entomology with a strong emphasis on taxonomy. Some review articles.

369   Annales de l'Institut Phytopathologique Benaki. 1957–. 1 per year. En, Fr. Institut Phytopathologique Benaki, rue Delta 8, Kiphissia, Athens, Greece.

Original research on phytopathological topics in Greece, including 40% agricultural entomology.

370   Annales de Parasitologie Humaine et Comparée. 1923–. 6 per year. Fr, En. Masson, 120 boulevard St Germain, F-75280 Paris Cedex 06, France.

Original research on human and comparative parasitology including 20% medical entomology.

371   Annales de la Société Entomologique de France. 1832–. 4 per year. Fr, En. Société Entomologique de France, 45 rue Buffon, F-75005 Paris, France.

Original research on general and applied entomology. Very occasional book reviews.

372   Annals of Applied Biology (and supplements). 1914–. 9 per year. En. Biochemical Society Book Depot, Box 32, Commerce Way, Whitehall Industrial Estate, Colchester, Essex CO2 8HP, England.

The annals of the Association of Applied Biologists. Original research on applied biology, including 10% applied entomology. Author index but no subject index for each volume.

373 **Annals of the Entomological Society of America.** 1908–. 6 per year. En. Entomological Society of America, 4603 Calvert Road, College Park, Maryland 20740, USA.

Original research. Papers do not have economic application or chemical control as their primary objective. Author index. Occasional book reviews, and obituaries.

374 **Annals and Magazine of Natural History** (including **Annals of Natural History**). 1838–1967. 12 per year. En.

This title has been continued as *Journal of Natural History* (entry 497). Many of the early volumes of the magazine contain important papers on entomology still often referred to as well as papers of historic importance.

375 **Annals of Tropical Medicine and Hygiene.** 1907–. 6 per year. En. Academic Press, 24–28 Oval Road, London NW1 7DX, England.

Some 10% medical entomology. In some years the entomological content has gone down, but it does vary from volume to volume. Annual author and contents list.

376 **Annual Review of Entomology.** 1956–. 1 per year. En. Annual Reviews, 4139 El Camino Way, Palo Alto, California 94306, USA.

Invited critical articles by qualified authors, reviewing significant developments within major disciplines. Author and subject indexes. Occasional cumulative subject indexes. Excellent bibliographies.

377 **Applied Entomology and Zoology.** 1966–. 4 per year. En, Fr, De. Japanese Society of Applied Entomology and Zoology, c/o Japan Plant Protection Association, Komagone, Toshima-ku, Tokyo 170, Japan.

Original research and short communications on applied zoology, applied entomology, agricultural chemicals and pest control equipment. Includes almost 100% economic entomology. Papers in Japanese are published in the *Japanese Journal of Applied Entomology and Zoology*.

378 **Aquatic Insects.** 1979–. 4 per year. En, Fr, De. Swets & Zeitlinger BV, Box 825, Lisse, NL-2160, Netherlands.

Taxonomy and ecology of aquatic insects.

379 **Atalanta (Zeitschrift der Deutschen Forschungszentrale für Schmetterlingswanderungen).** 1964–. 4 per year. De. Gesellschaft zur Förderung der Erforschung von Insektenwanderungen eV, München, Federal Republic of Germany.

Migration and distribution of Lepidoptera.

380 **Atalanta Norvegica (Norsk Lepidopterologisk Selskaps Tidsskrift).** 1968–. 2 per year. No, En. Zoologisk Museum, Sarsgate 1, Oslo 5, Norway.

Original research, taxonomy and field studies, mostly of the Norwegian fauna.

381 Australian Journal of Zoology (and supplements). 1953–. 6 per year. En. CSIRO, 314 Albert Street, East Melbourne, Victoria 3002, Australia.

Original research in all branches of zoology (anatomy, physiology, development, genetics, ecology and taxonomy). Includes 20% entomology. Annual author/subject indexes. Supplementary series consists chiefly of taxonomic revisions.

382 Bee World. 1919–. 4 per year. En. International Bee Research Association, Hill House, Gerrards Cross, Buckinghamshire SL9 0NR, England.

Original research on bees and beekeeping. Includes advertisements, book reviews and IBRA information.

383 Beiträge zur Entomologie. 1951–. 2 per year. De, En. Akademie-Verlag, Leipziger Str. 3/4, DDR-1080 Berlin, German Democratic Republic.

Original research on insect taxonomy. Bibliographies. Includes papers on Acarina. Carries extensive book reviews and papers on bibliographic research.

384 Biocontrol News and Information. 1980–. 4 per year. En. Commonwealth Institute of Biological Control, c/o 56 Queen's Gate, London SW7 5JR, England.

Abstracts, review articles, original research and news items relating to all aspects of biological control. Abstracts largely drawn from *Review of Applied Entomology* and other CAB abstracting journals, but supplemented by entries for semi-published material.

385 Biodeterioration Research Titles. 1966–. 4 per year. En. Biodeterioration Centre, University of Aston in Birmingham, St Peter's College, College Road, Saltley, Birmingham B8 3TE, England.

Bibliography of titles on most aspects of biodeterioration.

386 Bitki Koruma Bülteni. 1952–. 4 per year. Tu, En. Regional Plant Protection Institute, Kalaba, Ankara, Turkey.

Mainly original research on plant protection in Turkey. Published for the Turkish Ministry of Food, Agriculture and Animal Husbandry. Includes 45% agricultural entomology.

387 Biuletyn Instytutu Ochrony Roslin. 1957–. 1 per year. Pl, Ru, En. Instytut Ochrony Roslin, ul. Miczurina 20, 60-318 Poznán, Poland.

Annual reports on the more important pests and diseases of cultivated plants in Poland. Includes 80% agricultural entomology.

388   Boletin de la Asociacion Española de Entomología. 1977–. 1 per year. Es, En. Departamento de Zoología, Facultad de Ciencias, Universidad de Salamanca, Salamanca, Spain.

Research papers on the entomological fauna of Spain and its islands. Includes Acarina.

389   Bollettino dell'Istituto di Entomologia della Università degli Studi di Bologna. 1928–. 1 per year. It, En. Istituto di Entomologia, Università di Bologna, Bologna, Italy.

Original research papers mostly on the fauna of Italian territories. Some papers on agricultural entomology.

390   Bollettino del Laboratorio di Entomologia Agraria 'Filippo Silvestri', Portici. 1948–. 1 per year. It, En. Istituto di Entomologia Agraria dell'Università di Napoli, via Università 100, I-80055 Portici (NA), Italy.

Original contributions on research on general and applied entomology from the publishing institute. Contributions are accepted from foreign contributors when they coincide with the work of the Institute.

391   Bollettino Società Entomologica Italiana. 1869–. 4 per year. It, En, Fr. Società Entomologica Italiana, via Brigata Liguria 6, Genova, Italy.

Original research on most aspects of entomology. Includes the business of the Society. Occasional lists of members. Issue dates very late.

392   Bulletin et Annales de la Société Royale Belge d'Entomologie. 1857–. 4 per year. Fr, En. Société Royale Belge d'Entomologie, rue Vautier 31, B-1040 Bruxelles, Belgium.

Systematic entomology for the most part. Includes Acarina. Business of the Society is included in the first part of each issue, including obituaries of Fellows of the Society.

393   Bulletin of the British Museum (Natural History), Entomology Series. 1950–. Irregular. En, Fr, De. British Museum (Natural History), Department of Entomology, Cromwell Road, London SW7 5BD, England.

Original research based on the collections of the British Museum (Natural History). Author and contents list.

394   Bulletin of the California Insect Survey. 1950–. Irregular. En. University of California Press, 2223 Fulton Street, Berkeley, California 94720, USA.

Faunistic surveys, descriptions of and keys to the insects of California. 100% entomology.

395 Bulletin of Endemic Diseases. 1954–. 1 per year. En. Directorate of Endemic Diseases Institute, Baghdad, Iraq.

Original papers on tropical medicine and hygiene with emphasis on the epidemiology of parasitic and insect-borne diseases in the Middle East. Includes 35% medical entomology.

396 Bulletin of Entomological Research. 1910–. 4 per year. En. Commonwealth Institute of Entomology, 56 Queen's Gate, London SW7 5JR, England.

Original research and occasional review articles on arthropods of economic importance.

397 Bulletin of the Entomological Society of America. 1955–. 4 per year. En. Entomological Society of America, 4603 Calvert Road, College Park, Maryland 20740, USA.

Publishes the business of the Society, lists of members, plus articles of a more general nature. Carries extensive book reviews, bibliographies and obituaries, and proceedings of meetings.

398 Bulletin of the Entomological Society of Canada. 1969–. 4 per year. En. c/o Department of Biology, McMaster University, Hamilton, Ontario, Canada.

Reports and news of the work of the Society, and its meetings. Book reviews and biographical notes.

399 Bulletin of Grain Technology. 1963–. 3 per year. En. Foodgrain Technologists' Research Association of India, PO Box 10, Hapur, Uttar Pradesh, India.

Original research, reviews, scientific notes, news items and abstracts on grain technology with emphasis on storage aspects. Includes 70% agricultural entomology.

400 Bulletin of the National Institute of Agricultural Sciences, Series C (Plant Pathology and Entomology). 1952–. 1 per year. Ja, En. National Institute of Agricultural Sciences, Kannoudai, Yatabe, Tsukuba-gun, Ibaraki 305, Japan.

Original research on plant pathology, including 30% agricultural entomology.

401 Bulletin de la Société Entomologique d'Egypte. 1908–. 1 per year. En, Fr. Entomological Society of Egypt, PO Box 430, Cairo, Egypt.

Original research mostly on the Egyptian entomological fauna. About 2% of papers on Acarina.

402 Bulletin de la Société Entomologique de France. 1896–. (1832–1895 published with the *Annales* . . .). 10 per year. Fr, En, De. Société Entomologique de France, 45 rue de Buffon, F-75005 Paris, France.

Original research on systematic entomology and the entomological fauna of France. Carries information on the Society's business. Obituaries and biographies.

403 Bulletin de la Société de Pathologie Exotique et de Ses Filiales. 1908–. 6 per year. Fr, En, De. Masson, 120 boulevard St Germain, F-75280 Paris Cedex 06, France.

Mostly original research on tropical medicine, but includes 10% medical entomology.

404 Bulletin of the Society of Vector Ecologists. 1974–. 1 per year. En. Society of Vector Ecologists, 3827 West Chapman Avenue, Orange, California 92668, USA.

Original research on disease vectors and their control. Almost 100% medical entomology.

405 Bulletin SROP (WPRS Bulletin). 1974–. Irregular. En, Fr. Secrétariat Général de la SROP, Instituut voor Plantenziektenkundig Onderzoek, Binnenhaven 12, Wageningen, Netherlands.

The bulletin of the International Organization for Biological Control of Noxious Animals and Plants, West Palaearctic Regional Section. Each issue usually comprises papers read at IOBC meetings, but issues are sometimes given over to annual reports or, more rarely, to a single research paper. Largely agricultural entomology.

406 Bulletin of the Tohoku Agricultural Experiment Station. 1946–. Irregular. Ja, En. Tohoku National Agricultural Experiment Station, Morioka, Wate 020-01, Japan.

Original research carried out at Tohoku, including 20% economic entomology (mostly agricultural, some veterinary).

407 Bulletin of Zoological Nomenclature. 1943–. 4 per year. En. International Trust for Zoological Nomenclature, c/o British Museum (Natural History), Cromwell Road, London SW7 5BD, England.

Opinions and declarations of the International Commission for Zoological Nomenclature. Papers commenting on, supporting and opposing cases put before the commission for possible nomenclature changes.

Butterflies and Moths – *see* **Tyo to Ga**

408 Cahiers ORSTOM, Série Entomologie Médicale et Parasitologie. 1970–. 4 per year. Fr, En. Service des Publications de l'ORSTOM, 70–74 route d'Aulnay, F-93140 Bondy, France.

Original research papers on the systematics and biology of arthropods of medical, veterinary and parasitological interest, also papers on their control.

409 Cahiers de la Recherche Agronomique. 1948–. Irregular. En, Fr. Services d'Edition, d'Impression et de Diffusion, Institut National de la Recherche Agronomique, BP 415, Rabat, Morocco.

Primary agricultural research, including 60% agricultural entomology. Often published in the form of large monographs, particularly on economically important groups of insects.

410   California Agriculture. 1946–. 12 per year. En. California Agriculture, 317 University Hall, 2200 University Avenue, Berkeley, California 94720, USA.

Reports of progress in research by the Agricultural Experiment Station and Cooperative Extension, Division of Agricultural Sciences, University of California. 20% agricultural entomology (mostly pest control). Well illustrated with many colour photographs.

411   Canadian Entomologist. 1868–. 12 per year. En, Fr. Entomological Society of Canada, 1320 Carling Avenue, Ottawa, Ontario K1Z 7K9, Canada.

Original research. Some 80% systematics. The remainder economic entomology. Short notes.

412   Canadian Journal of Zoology. 1951–. 12 per year. En, Fr. National Research Council of Canada, Ottawa, Ontario K1A 0R6, Canada.

Original articles, notes and reviews in all fields of zoology. Includes 10% entomology (mainly applied aspects).

413   Cecidologica Indica. 1966–. 3 per year. En. Cecidological Society of India, 14 Park Road, Allahabad 211002, India.

Devoted entirely to cecidological studies.

414   Coleopterist's Bulletin. 1947–. 4 per year. En. c/o Department of Entomology, Stop 168, Smithsonian Institution, Washington, DC 20560, USA.

Biology, distribution and systematics of Coleoptera.

415   Contributions of the American Entomological Institute. 1964–. Irregular. En. American Entomological Institute, 5950 Warren Road, Ann Arbor, Michigan 48105, USA.

Series of monographs, mostly devoted to taxonomy. In recent years, the papers have mostly covered medical entomology, and mosquitoes in particular.

416   Coton et Fibres Tropicales. 1946–. 4 per year. En, Es, Fr. Institut de Recherches du Coton et des Textiles Exotiques, 34 rue des Renaudes, F-75017 Paris, France.

Original research on all aspects of cotton and tropical fibres, including 25% agricultural entomology.

417   Current Research. 1972–. 12 per year. En. University of Agricultural Sciences, Hebbal, Bangalore 560024, India.

Research news and correspondence (all brief) on all aspects of agricultural science, including 13% economic entomology (mainly agricultural).

418   Data Sheets on Quarantine Organisms. 1977–. Irregular. En, Fr. Organisation Européenne et Méditerranéenne pour la Protection des Plantes, 1 rue Le Notre, F-75016 Paris, France.

Each 'sheet' (usually about 4 pages in length) covers a single pest or pathogen and gives data on: nomenclature, biology, host plants, distribution, identification, economic importance, potential in the OEPP region, means of entry, inspection and treatment methods, and references. Each is illustrated by plates, usually in colour, showing the organism and/or the type of damage caused. Issued in various series, divided into two lists: A1, covering organisms not present in the OEPP region, and A2, covering those which are present in the region.

419 Deutsche Entomologische Zeitschrift (Berliner Entomologische Zeitschrift). 1857– (N.S. and renumbered 1954–). 5 per year. En, Fr, De. Akademie-Verlag, Leipzigerstr. 3–4, DDR-108 Berlin, German Democratic Republic.

Original research on all fields of entomology, but an emphasis on systematics. Includes Acarina.

420 Distribution Maps of Pests, Series A (Agricultural). 1951–. 18 per year. En. Commonwealth Institute of Entomology, 56 Queen's Gate, London SW7 5JR, England.

Eighteen new or revised maps are issued each year, each map showing the world distribution of a single agricultural insect pest. The reverse of each map lists supporting bibliographic references. Over 400 maps have been issued so far. An index is issued at irregular intervals.

421 East African Agricultural and Forestry Journal. 1960–. Irregular. En. East African Agricultural and Forestry Journal, PO Box 30148, Nairobi, Kenya.

Original research in the fields of agriculture, veterinary science and forestry and closely allied subjects. Includes 15% economic entomology. Book reviews.

422 Ecological Entomology. 1976–. 4 per year. En. Blackwell Scientific, PO Box 88, Oxford, England.

Original research papers on field biology and natural history of terrestrial and aquatic insects; descriptions of ecological methods and apparatus; integrated control of pest populations, ecological aspects of insect archaeology.

423 Entomologia Experimentalis et Applicata. 1958–. 6 per year. En, Fr, De. EEA Administration, c/o Bibliotheek Nederlandse Vereniging, Plantage Middenlaan 64, NL-1018 DH Amsterdam, Netherlands.

Mostly original research on experimental biology and ecology (pure and applied) of insects and some other land arthropods (e.g. mites). Biased towards agricultural rather than medical or veterinary entomology.

424 Entomologia Generalis (Entomologica Germanica). 1974–. 4 per year. De, En, Fr. Gustav Fischer Verlag, Wollgrasweg 49, Postfach 720143, D-7000 Stuttgart 72, Federal Republic of Germany.

An 'international journal for scientific entomology'. Original research on physiology, ecology, ethology, morphology and phylogenetics. Carries book reviews and biographies. Annual index.

425   Entomologica Basiliensia. 1975–. Irregular. En. Naturhistorisches Museum, Entomologisches Abteilung, Basel, Switzerland.

Papers on insects collected on the Bhutan Expedition 1972 des Naturhistorischen Museums in Basel. Additional papers on original research on general entomology.

426   Entomologica Scandinavica. 1970–. 4 per year. En. Swedish Natural Science Foundation, Wenner-Gren Center, Box 23136, S-104 35 Stockholm, Sweden.

Original papers on systematics and taxonomic entomology of international interest, but priority is given to papers on the Holarctic fauna. Supplements at irregular intervals.

427   Entomological Research Bulletin, Korea. 1965–. Irregular. Ko, En. Korean Entomological Institute, Korea University, Seoul 132, Korea.

100% Korean entomology (mainly applied).

428   Entomological Review, Washington. 1957–. 4 per year. En. Scripta Publishing, 7061 Eastern Avenue, Silver Spring, Maryland 20910, USA.

Early volumes were a complete cover-to-cover translation of *Entomologicheskoe Obozrenie* (entry 429), but more recent volumes include only the scientific papers, omitting books reviews, obituaries and business of the Russian Entomological Society.

429   Entomologicheskoe Obozrenie. 1901–. 4 per year. Ru. Academy of Sciences, Entomology Section, Leningrad, USSR.

General entomology, but mostly of the Palaearctic fauna. Carries papers on Acarina. Good book reviews, biographies and bibliographies. Information on the Russian Entomological Society. *See also* entry 428 (translation).

430   Entomologie et Phytopathologie Appliquées. 1946–. Irregular. Pe, En. Institut de Recherches Entomologiques et Phytopathologiques, PO Box 3178, Tehran, Iran.

Original Iranian research on agricultural entomology (80%) and phytopathology.

431   Entomologische Arbeiten aus dem Museum G. Frey. 1950–. Irregular. De, Fr, En. Museum G. Frey, Entomologisches Institut, Tutzing bei München, Federal Republic of Germany.

Systematics of Coleoptera only.

432   Entomologische Berichte. 1963–. 2 per year. De. Kulturbund der DDR, Zentraler Fachauschuss Entomologie beim Zentralvorstand der Gesellschaft für Natur and Umwelt, Hessische Str. 11–12, DDR-1040 Berlin, German Democratic Republic.

Original research on systematics, biology and ecology. Carries book reviews, biographies, obituaries and bibliographies.

433 Entomologische Berichten. 1901–. 12 per year. Nd, En, Fr, De. B.J. Lempke, Plantage Middenlaan 64, NL-1018 DH Amsterdam, Netherlands.

General entomology, but emphasis on the Dutch fauna. Book reviews and biographies. Information from the Dutch Entomological Society.

434 Entomologische Blätter für Biologie und Systematik der Käfer. 1905–. 2 per year. De. Goecke & Evers, Dürerstr. 13, D-4150 Krefeld, Federal Republic of Germany.

Original research on the biology and systematics of Coleoptera. Book reviews and bibliographies.

435 Entomologische Mitteilungen aus dem Zoologischen Museum Hamburg. 1952–. Irregular. De. Zoologisches Institut und Zoologisches Museum der Universität Hamburg, Krause-Druck, D-2160 Stade, Federal Republic of Germany.

Original research on systematics, biology and physiology. Biographies included.

436 Entomologische Nachrichten. 1957–. 12 per year. De. Dr B. Klausnitzer, Lannerstr. 5, DDR-8020 Dresden, German Democratic Republic.

Biology and systematics of all groups of insects.

437 Entomologische Zeitschrift. 1887–. 12 per year. De. Internationales Entomologisches Verein, Frankfurt am Main, Federal Republic of Germany.

Short articles on most areas of the subject, with a slight emphasis on Lepidoptera. Occasional book reviews and biographies.

438 Entomologisk Tidskrift. 1880–. 3 per year. Sw, En. Sveriges Entomologiska Förening, Zoologiska Institutionen, Helgonävagen 3, S-223 62 Lund, Sweden.

Original research in all fields of entomology. Short notes, book reviews and biographical material. Annual index.

439 Entomologiske Meddelelser. 1887–. 3 per year. Da, En. Entomological Society of Copenhagen, Zoologisk Museum, Universitetsparken 151, DK-2100 København, Denmark.

Original research and review articles, chiefly in Danish. Most aspects of international entomology. Biographies and book reviews.

440 Entomologiste. 1944–. 6 per year. Fr. A. Villiers, 45 bis rue de Buffon, F-75005 Paris, France.

Short articles on most fields of entomology. Aimed more particularly at the informed amateur. Annual index.

441 Entomologist's Gazette. 1950–. 4 per year. En. E.W. Classey, PO Box 93, Faringdon, Oxfordshire SN7 7DL, England.

A 'journal of Palaearctic entomology'. Some emphasis on the Lepidoptera – original research papers. Book reviews.

442   Entomologist's Monthly Magazine. 1865–. 12 per year. En. Entomologist's Monthly Magazine Ltd, 7 Thorncliffe Road, Oxford OX2 7BA, England.

Publishing dates have slipped badly, more frequently published quarterly. Subject area not restricted to British entomology. Original papers on biology and systematics; publishes book reviews, bibliographies and biographies. Many short notes.

443   Entomologist's Record and Journal of Variation. 1890–. 12 per year. En. J.M. Chalmers-Hunt, St Teresa, 1 Harcourts Close, West Wickham, Kent, England.

Mostly short papers on most aspects of entomology, a strong bias towards Lepidoptera. Short notes and communications, book reviews and biographical notices. Annual author and subject index. A supplement in parts, some 10–20 pages per issue, *Lepidoptera of Kent*. Volume 3 is now complete.

444   Entomology Circular. 1962–. Irregular. En. Division of Plant Industry, Florida Department of Agriculture and Consumer Services, Box 1269, Gainesville, Florida 32602, USA.

Each issue normally gives a concise review of the biology, ecology and control of a particular insect pest species or group that is of agricultural importance in Florida. As many of the pests of Florida are cosmotropical in distribution, these notes have much wider application than to Florida alone.

445   Entomology Newsletter. 1974–. Irregular. En. Central South Experiment Station, IAA/PLANALSUCAR, Caixa Postal 158, 13.600 Araras, SP, Brazil.

The newsletter of the Entomology Section of the International Society of Sugarcane Technologists. Devoted to original research, brief notes and news items on all aspects of sugarcane entomology.

446   Entomon. 1976–. 4 per year. En. Entomon, Department of Zoology, University of Kerala, Kariavattom, Trivandrum 695581, India.

Original research, brief communications, reports and new records on all aspects of entomology excluding gross anatomy, histology and morphology. Emphasis on agricultural entomology and pest control.

447   Entomophaga. 1956–. 4 per year. En, Fr. Balthazar Publications, 9 rue Edonard-Jacques, F-75014 Paris, France.

Devoted to original research on biological and integrated control, this is the official journal of the International Organization for Biological Control of Noxious Plants and Animals. Almost exclusively entomological. Also includes occasional book reviews.

448   Environmental Entomology. 1972–. 6 per year. En. Entomological Society of America, 4603 Calvert Road, College Park, Maryland 20740, USA.

Original research on the behaviour of insects and their interaction with the biological, chemical and physical constituents of the environment.

449 EOS (Revista Española de Entomología). 1969–. Irregular. Es, En. Instituto Español de Entomología, José Gutiérrez Abascal 2, Madrid 6, Spain.

Original research papers on all families of insects, and on Acari. Nominally annual volumes, but appears late and irregularly.

450 Experientia. 1945–. 12 per year. En, Fr, De, It. Fr. Birkhäuser Verlag, Elisabethenstr. 19, CH-4010 Basel, Switzerland.

A monthly journal of pure and applied science, including a high proportion of papers on economic entomology.

451 FAO Animal Production and Health Papers. 1977–. Irregular. En. FAO, via delle Terme di Caracalla, I-00100 Rome, Italy.

Each issue deals with a single topic. Coverage is 25% veterinary entomology.

452 Fauna Norvegica, Series B (Norwegian Journal of Entomology). 1920–. 6 per year. En, No. Norsk Zoologisk Tidsskriftsentral, Zoologisk Museum, Sarsgate 1, Oslo 5, Norway.

Mostly faunistic and zoogeographical papers. Checklists, faunal lists, type catalogues and regional keys.

453 Fiji Agricultural Journal (Agricultural Journal, Department of Agriculture, Fiji). 1928–61; 1970–. 2 per year. En. Ministry of Agriculture and Fisheries, PO Box 358, Suva, Fiji.

Original research and reviews relevant to agriculture, fisheries and forestry in Fiji. Includes 10% economic entomology.

454 Florida Entomologist. 1920–. 4 per year. En. Florida Entomological Society, PO Box 12425, University Station, Gainesville, Florida 32604, USA.

Original research papers on general and economic entomology including papers on insecticides and pesticides. Papers on mites are included.

455 Folia Entomologica Hungarica. 1923–. 6 per year. Hu, En, De, Fr. Hungarian Natural History Museum, Baross u. 13, 1088 Budapest, Hungary.

Original research on all aspects of entomology, slight emphasis on Palaearctic fauna. Includes papers on Arachnida.

456 Folia Entomológica Mexicana. 1961–. Irregular. Es, En. Sociedad Mexicana de Entomología, Apartado Postal 31-312, Mexico 7, DF, Mexico.

General entomology, but restricted to the Mexican fauna.

457 Folia Parasitologica. 1966–. 4 per year. En. Institute of Parasitology, Flemingovo nam 2, 166-32 Prague 6-Dejvice, Czechoslovakia.

Mainly original research on all aspects of parasitology, including 15% medical and veterinary entomology.

458   Fragmenta Entomologica. 1951–. Irregular. It, En. Università degli Studi di Roma, Istituto di Zoologia, Roma, Italy.

Covers all zoogeographical areas. 95% Insecta, but does carry some papers on Crustacea and other invertebrate groups.

459   General and Applied Entomology (Journal of the Entomological Society of Australia (NSW)). 1964–. 1 per year. En. Entomological Society of Australia (NSW), PO Box 22, Five Dock, New South Wales 2046, Australia.

Original research papers on all aspects of entomology, with emphasis on economically important species. Book reviews.

460   Ghana Journal of Agricultural Science. 1968–. 3 per year. En. Ghana University Press, PO Box 4219, Accra, Ghana.

Intended as an outlet for papers concerning West African agriculture and related disciplines. Each issue divided into 3 sections: experimental and environmental sciences, research and development notes, and documentation. Papers published in the first section are also issued as preprints to interested West African institutions. Includes 15% agricultural entomology. Book reviews.

461   Gradinarska i Lozarska Nauka. 1964–. 8 per year. Bu, Ru, En, Fr. Akademiia na Selskostopanskite Nauki, Ul. Akad. G. Bonchev, 1113 Sofia, Bulgaria.

Bulgarian journal of horticulture and viticultural science. Includes a small percentage of papers on insect pests.

462   Indian Journal of Agricultural Sciences. 1931–. 12 per year. En. Indian Council of Agricultural Research, Krishi Bhavan, New Delhi 110001, India.

Original research, short research notes and critical reviews on agricultural sciences, including 15% economic entomology (mainly agricultural).

463   Indian Journal of Entomology. 1939–. 4 per year. En. Entomological Society of India, Indian Agricultural Research Institute, Division of Entomology, New Delhi 110012, India.

Original papers on all aspects of entomology, including economic entomology, carries some review articles.

464   Indian Journal of Plant Protection. 1973–. 2 per year. En. Plant Protection Association of India, Hyderabad, AP, India.

Original research on all aspects of plant and soil sciences, including 10% agricultural entomology.

465   Insect Biochemistry. 1971–. 6 per year. En. Pergamon Press, Headington Hill Hall, Oxford OX3 0BW, England.

Original research and reviews on all aspects of insect biochemistry. Almost 100% economic entomology. Annual author index but no subject index.

466 Insect Science and Its Application. 1980–. 4 per year. En, Fr. Pergamon Press, Headington Hill Hall, Oxford OX3 0BW, England.

Original research and reviews on all aspects of tropical insects and related arthropods, and the application of new discoveries to fields such as pest and vector management, and the use of beneficial insects.

467 Insecta Matsumurana (and supplements). 1926–. 4 per year (becoming very irregular). En. Insecta Matsumurana, Entomological Institute, Faculty of Agriculture, Hokkaido University, Sapporo 060, Japan.

Original research on the systematics and biology of insects.

468 Insectes Sociaux (International Journal for the Study of Social Arthropods). 1954–. 4 per year. Fr, En, De. Masson, 120 boulevard St Germain, Paris, France.

Original research on the biology, morphology and systematics of social insects. Book reviews.

469 Insecticide and Acaricide Tests. 1976–. 1 per year. En. Entomological Society of America, 4603 Calvert Road, College Park, Maryland 20740, USA.

Each issue contains several hundred submitted reports of insecticide and acaricide tests, together with a full index, a section giving information on composition, formulation and supplies of the products listed, and details of new products.

470 International Biodeterioration Bulletin. 1965–. 4 per year. En. Biodeterioration Centre, University of Aston in Birmingham, St Peter's College, Saltley, Birmingham B8 3TE, England.

Original works and review articles on all aspects of biodeterioration.

471 International Journal of Insect Morphology and Embryology. 1971–. 6 per year. En. Pergamon Press, Headington Hill Hall, Oxford OX3 0BW, England.

Original contributions on all aspects of gross morphology, palaeomorphology, macro- and microanatomy and ultrastructures of insects. Related arthropods which have a direct bearing on the understanding of insect morphology.

472 International Journal of Invertebrate Reproduction. 1979–. 6 per year. En, Fr. Elsevier/North-Holland, Journal Division, PO Box 211, NL-1000 AE Amsterdam, Netherlands.

Original research and short reviews on all aspects of the reproduction of the Invertebrata (excluding Protozoa). Includes 10% economic entomology. Annual subject index but no author index.

473 International Pest Control. 1962–. 6 per year. En. McDonald Publications, 268 High Street, Uxbridge, Middlesex UB8 1UA, England.

Original research, news items, advertisements and book reviews and information on patents relevant to crop and stock protection, public health and wood preservation. About 60% economic entomology, mainly agricultural.

474   International Rice Research Newsletter. 1976–. 6 per year. En. International Rice Research Institute, PO Box 933, Manila, Philippines.

Brief original research and progress reports on all aspects of rice cultivation including 40% agricultural entomology.

475   Iranian Journal of Public Health. 1972–. 4 per year. En. Iranian Journal of Public Health, 16 Avenue Azar, PO Box 1310, Tehran, Iran.

Original research on and reviews of public health topics in Iran. Includes 10% medical entomology.

476   Israel Journal of Entomology. 1966–. Irregular. En. Entomological Society of Israel, c/o Volcani Institute for Agricultural Research, Bet-Dagan, PO Box 6, Israel.

Devoted to all aspects of entomology. Some emphasis on economic entomology in recent issues.

477   Japanese Journal of Applied Entomology and Zoology. 1957–. 4 per year. Ja, En. Agricultural Society of Applied Entomology and Zoology, c/o Japan Plant Protection Association, Komagome, Toshima-ku, Tokyo 170, Japan.

Original research and short communications on applied entomology (almost 100%) and applied zoology. Papers in English published in *Applied Entomology and Zoology* (entry 377).

478   Japanese Journal of Sanitary Zoology. 1950–. 4 per year. En, Ja. Japan Society of Sanitary Zoology, c/o Institute of Medical Science, University of Tokyo, Shiroganedai, Minatoku, Tokyo, Japan.

Almost exclusively original research papers and short notes on medical entomology.

479   Journal of Agriculture, Tasmania (Tasmanian Journal of Agriculture). 1929–. 4 per year. En. Tasmanian Department of Agriculture, 1 Franklin Wharf, PO Box 192B, Hobart, Tasmania 7001, Australia.

Some original research but mainly extension-type papers, of which about 10% are on or relevant to economic entomology.

480   Journal of Animal Ecology. 1932–. 3 per year. En. Blackwell Scientific, Osney Mead, Oxford OX2 0EL, England.

The journal of the British Ecological Society, devoted to original research on all aspects of animal ecology, with emphasis on the experimental or theoretical approach. Includes 50% applied entomology.

481   Journal of Apicultural Research. 1962–. 4 per year. En. International Bee Research Association, Hill House, Gerrards Cross, Buckinghamshire SL9 0NR, England.

Original research and reviews on bees and related subjects. 100% apiculture. Annual author index but no subject index.

482 Journal of Applied Ecology. 1964–. 3 per year. En. Blackwell Scientific, Osney Mead, Oxford OX2 0EL, England.

Original research and reviews on all aspects of applied ecology. Includes 8% economic entomology. Book reviews. Annual author index but no subject index.

483 Journal of the Australian Entomological Society (Journal of the Entomological Society of Queensland). 1962–. 4 per year. En. Australian Entomological Society, Department of Primary Industries, Meiers Road, Indooroopilly, Queensland 4068, Australia.

All aspects of entomological research with particular reference to the Australian fauna.

484 Journal of Chemical Ecology. 1975–. 6 per year. En. Plenum, 227 West 17th Street, New York, NY 10011, USA.

Original research and reviews on the interactions of organisms with their environment mediated by the chemicals they produce. Includes 75% applied entomology (mainly chemistry and application of pheromones). Annual author/subject indexes.

485 Journal of Communicable Diseases. 1969–. Irregular. En. Indian Society for Malaria and Other Communicable Diseases, 22 Sham Nath Marg, Delhi 110054, India.

Original Indian research on all aspects of communicable diseases, including insect vectors. Includes 45% medical entomology (mainly mosquito biology, ecology and control).

486 Journal of Economic Entomology. 1908–. 6 per year. En. Entomological Society of America, 4603 Calvert Road, College Park, Maryland 20740, USA.

Original research on all aspects of economic entomology (agricultural, medical and veterinary); adverts. Annual author/subject indexes. 100% economic entomology.

487 Journal of Entomological Research. 1977–. 2 per year. En. Malhotra Publishing House, A 38/3, Mayapuri Industrial Area, New Delhi 110064, India.

Original contributions on all fundamental and applied entomology.

488 Journal of the Entomological Society of Southern Africa. 1939–. 2 per year. En. Entomological Society of Southern Africa, PO Box 103, Pretoria, South Africa.

General entomology, some papers on Arachnida, mostly of South African fauna. Book reviews, biographical notices.

489 Journal of Experimental Biology. 1930–. 6 per year. En. Cambridge

University Press, The Edinburgh Building, Shaftesbury Road, Cambridge CB2 2RU, England.

Original research and short communications on all aspects of experimental biology, including 15% economic entomology. Separate annual author and subject indexes.

490   Journal of the Faculty of Agriculture, Kyushu University. 1948–. Irregular. En. Faculty of Agriculture, Kyushu University, Fukuoka, Japan.

Original research carried out within Kyushu University. 10% agricultural entomology.

491   Journal of the Georgia Entomological Society. 1966–. 4 per year. En. Georgia Entomological Society, USDA, PO Box 22909, Savannah, Georgia 31403, USA.

Papers largely on economic entomology, agricultural pests and their control.

492   Journal of Insect Physiology. 1957–. 12 per year. En, De, Fr. Pergamon Press, Headington Hill Hall, Oxford OX3 0BW, England.

Research papers in the field of insect physiology and closely related problems of other arthropods if of general interest.

493   Journal of Invertebrate Pathology. 1965–. 6 per year. En. Academic Press, 111 Fifth Avenue, New York, NY 10003, USA.

Original research concerned with the nature and study of microbial and amicrobial diseases of invertebrates, the suppression of these diseases in beneficial invertebrates, their use in the control of invertebrate pests, and those phases of general invertebrate microbiology that may have some relation to disease in these animals. Annual author/subject indexes. Adverts.

494   Journal of the Kansas Entomological Society. 1928–. 4 per year. En. Kansas Entomological Society, PO Box 368, Lawrence, Kansas 66044, USA.

Original research on all aspects of entomology including pest control. Short notes, annual index.

495   Journal of the Lepidopterists' Society (Lepidopterist's News). 1947–. 4 per year. En. Lepidopterists' Society, F.C. Chew, Department of Biology, Tufts University, Medford, Massachusetts 02155, USA.

Original research on the systematics, ecology and biology of the Lepidoptera. Book reviews, biographies and bibliographies.

496   Journal of Medical Entomology. 1964–. 6 per cent. En. Department of Entomology, Bishop Museum, 1355 Kalihi Street, Honolulu, Hawaii 96819, USA.

Original research on all aspects of medical entomology, including systematics of insects and other arthropods of public health or veterinary significance. Book reviews. 100% medical/veterinary entomology.

497   Journal of Natural History. 1967– (preceded by *Annals and Magazine of Natural History*). 6 per year. En. Taylor & Francis, 4 John Street, London WC1N 2ET, England.

Biology and systematics; book reviews; original research, biology and systematics of zoological groups including entomology. Some 30% of papers are entomological.

498   Journal of the New York Entomological Society. 1893–. 4 per year. En. New York Entomological Society, Waksman Institute of Microbiology, Rutgers University, New Brunswick, New Jersey 08903, USA.

Original papers on all areas of entomology.

499   Journal of Parasitology. 1914–. 6 per year. En. American Society of Parasitologists, 1041 New Hampshire Street, Lawrence, Kansas 66044, USA.

Original research on mainly animal parasitology, including 10% medical and veterinary entomology. Book reviews, announcements.

500   Journal of Research on the Lepidoptera. 1962–. 4 per year. En. Lepidoptera Research Foundation, 2559 Puesta del Sol Road, Santa Barbara, California 93105, USA.

General Lepidoptera research. Book reviews, bibliographies, biographies and short notes.

501   Journal of Stored Products Research. 1965–. 4 per year. En. Pergamon Press, Headington Hill Hall, Oxford OX3 0BW, England.

Original research on insects, mites, fungi and other organisms associated with stored products, or on aspects of the physical and chemical nature of the stored product environment which have relevance to these organisms and their control. Occasional reviews and short communications. Almost 100% economic entomology. Annual author index but no subject index.

502   Khemoretseptsiya Nasekomykh. 1976–. 1 per year. Ru, En. Insect Chemoreception Laboratory, Institute of Zoology and Parasitology, Lithuanian Academy of Sciences, MTP-8, Ul. K. Poželos 54, 232600 Vilnius, Lithuanian SSR, USSR.

Soviet research journal on all aspects of insect chemoreception.

503   Koleopterologische Rundschau. 1912–. Irregular. De. Zoologisch-Botanische Gesellschaft, Forstliche Bundesversuchsanstalt, A-1131 Wien, Austria.

Biology and systematics of Palaearctic Coleoptera.

504   Kontyû (Journal of the Entomological Society of Japan). 1926–. 4 per year. Ja, En, De. Entomological Society of Japan, c/o Department of Zoology, National Science Museum (Natural History), 3-23-1 Jjakunai Cho, Shinjuku-ku, Tokyo 160, Japan.

Taxonomy, morphology and ecology of Japanese insects (some 80% on the Japanese fauna). Book reviews and bibliographies.

505 Korean Journal of Entomology. 1971–. 2 per year. En, Ko. Entomological Society of Korea, Korean Entomological Institute, Korean University, 1 Ana-Dong, Sungbuk-ku, Seoul 132, Korea.

Original articles, mostly on the Korean fauna.

506 Korean Journal of Plant Protection. 1971–. 4 per year. Ch, En. Korean Society of Plant Protection, c/o Department of Agricultural Biology, College of Agriculture, Seoul National University, Suwon, 170, Korea.

Original research and reviews on Korean plant protection, including 15% agricultural entomology.

507 Lambillionea. 1926–. 6 per year. Fr, En. Union des Entomologistes Belges, 34 avenue du Manoir d'Anjou, B-1150 Bruxelles, Belgium.

Research papers on the entomological fauna of Belgium, but principally the Lepidoptera. Supplements issued irregularly.

508 Latvijas Entomologs. 1960–. Irregular. Lv, Ru. Latvijas Entomologi-jas Biedribas, Latvijas PSR, Zinatnu Akademijas, Riga, Latvia, USSR.

Original research on insect fauna with emphasis on pest species. Supplements published on an irregular basis.

509 Lepidoptera. 1946–51; 1965– (NS). 2 per year. Da. Lepidopterolo-gisk Forening, c/o Mr Carsten Hviid, Godsbanegade 31, 4tv., DK-1722 Denmark.

Biology, distribution and taxonomy of Danish Lepidoptera. Annual index.

510 Linneana Belgica. 1958–. 4 per year. Fr, En. Mr R. Leestmans, Parris Saint-Gilles 4, B-1060 Bruxelles, Belgium.

Original research on the Belgian insect fauna, and particularly the Lepidoptera.

511 Malaysian Agricultural Journal. 1965–. Irregular. En. Ministry of Agriculture and Cooperatives, Kuala Lumpur, Malaysia.

Reports of original research, investigations and extension work performed principally by the various divisions and bodies under the Ministry of Agriculture, Malaysia. Includes 30% agricultural entomology.

512 Malaysian Applied Biology. 1972. 2 per year. En. Malaysian Society of Applied Biology, c/o Faculty of Science, University of Malaya, Kuala Lumpur, Malaysia.

Original research in biology and related fields, including 10% economic entomology.

513 Meditsinskaya Parazitologiya i Parazitarnye Bolezni. 1932–. 6 per year. Ru, En. Mezhdunarodnaya Kniga (Distributor), Moscow G-200, USSR.

Original research, brief notes, news items and book reviews including 40% medical/veterinary entomology.

514 Memoirs of the American Entomological Institute. 1961–. Irregular. En. American Entomological Institute, 5950 Warren Road, Ann Arbor, Michigan 48105, USA.

Large monographs separately numbered. Mostly on Neotropical fauna, but not exclusively so.

515 Memoirs of the American Entomological Society. 1916–. Irregular. En. American Entomological Society, Academy of Natural Sciences, 1900 Race Street, Philadelphia, Pennsylvania 19103, USA.

Numbered monograph series. Taxonomic entomology.

516 Memoirs of the Entomological Society of Canada (previously, supplements to **Canadian Entomologist**). 1956–. Irregular.' En, Fr. Entomological Society of Canada, 1320 Carling Avenue, Ottawa, Ontario K1Z 7K9, Canada.

Numbered monographs; original research papers on taxonomic entomology.

517 Memorie della Società Entomologica Italiana. 1921–. Irregular. It. Dr E. Berio, Società Entomologica Italiana, via Brigata Liguria 9, Genova, Italy.

Original research and review papers on taxonomic entomology. Nominally an annual volume, but in practice issues are very irregular.

518 Miscellaneous Publications of the Entomological Society of America. 1959–. Irregular. En. Entomological Society of America, 4603 Calvert Road, College Park, Maryland 20740, USA.

All aspects of entomology. A high percentage of review-style articles with good bibliographies.

519 Miscellaneous Reports, Centre for Overseas Pest Research. 1971–. Irregular. En. Centre for Overseas Pest Research, College House, Wrights Lane, London W8 5SJ, England.

Each report deals with a single pest control topic which is usually relevant to applied entomology.

520 Mitteilungen aus der Biologischen Bundesanstalt für Land- und Forstwirtschaft. 1954–. Irregular. De, En. Kommissionsverlag Paul Parey, Lindenstr. 44–47, D-1000 Berlin 61, Federal Republic of Germany.

Each issue deals with a separate German plant protection topic, usualy original research but occasionally reviews, bibliographies or conference proceedings. Includes 60% agricultural entomology.

521 Mitteilungen der Entomologischen Gesellschaft Basel. 1951–. Irregular. De, Fr, En. Entomologische Gesellschaft Basel, Färberstr. 1, CH-4047 Basel, Switzerland.

Strong emphasis on Lepidoptera, but other groups included. Not restricted to Swiss fauna.

522 Mitteilungen der Münchner Entomologischen Gesellschaft. 1910–. 2 per year. De. Dr W. Forster, Münchner Entomologische Gesellschaft, Maria-Ward-Str. 1b, D-8000 München 19, Federal Republic of Germany.

All groups of insects, not restricted to any zoogeographical region. Carries extensive book reviews. Annual index.

523 Mitteilungen der Schweizerischen Entomologischen Gesellschaft. 1862–. 2 per year. De, Fr. Prof. Dr V. Delucchi, Institut für Phytomedizin, ETH-Zentrum, Clausiusstr. 21, CH-8092 Zürich, Switzerland.

Carries original papers on general, taxonomic and applied entomology, mostly on the Swiss fauna.

524 Mosquito News (Journal of the Mosquito Control Association). 1941–. 4 per year. En. American Mosquito Control Association, 5545 East Shields Avenue, Fresno, California 93727, USA.

Papers mostly on mosquito control. Each issue carries a classified current literature review. Carries book reviews and biographical material. Annual index.

525 Mosquito Systematics. 1972–. 4 per year. En. American Mosquito Control Association, 5545 East Shields Avenue, Fresno, California 93727, USA.

Original research papers on mosquito taxonomy. Book reviews, editorial notes.

526 Nachrichtenblatt der Bayerischen Entomologen. 1952–. 12 per year. En. Münchner Entomologische Gesellschaft, Maria-Ward-Str. 1b, D-8000 München 19, Federal Republic of Germany.

Original research on all branches of entomology. Normally published monthly but a cumulative volume is usually sent to subscribers.

527 Nasekomye Mongolii (Insects of Mongolia). 1972–. Irregular. Ru. Mr I. M. Kerzhner, Academy of Sciences of the USSR and Academy of Sciences of the MPR, Zoological Institute, Institute of General and Experimental Biology, Leningrad, USSR.

Volumes published irregularly. Contain papers on the joint Soviet–Mongolian complex biological expedition. Most of the published papers in these volumes are on the taxonomy of insects collected on that expedition. Issued as hardback bound volumes.

528 Netherlands Journal of Plant Pathology. 1963–. 6 per year. En, Nd. Nederlandse Planteziektenkundige Vereniging, PO Box 31, NL-6700 AA Wageningen, Netherlands.

Original research, short communications and book reviews on all aspects of plant pests and diseases, including 10% agricultural entomology.

529 New Entomologist. 1952–. Irregular. Ja, En. Entomological Society of Shinshu, c/o Laboratory of Biology, Faculty of Textile Science and Technology, Shinshu University, Ueda, Japan.

Original research papers, with strong emphasis on the Japanese Lepidoptera. Includes 60% of papers in Japanese, the remainder in English.

530 New Zealand Entomologist. 1951–. 1 per year. En. Distributions Secretary, Entomological Society of New Zealand, 6 Ocean View Road, Huia, New Zealand.

Original research and brief communications relating to entomology (largely economic) in New Zealand. Also includes Society news and business matters, and book reviews.

531 Nigerian Journal of Entomology. 1974–. Irregular. En. Dr F. Osuji, Dept of Agricultural Biology, University of Ibadan, Ibadan, Nigeria.

Any aspect of entomology in Nigeria or other tropical country, but a strong emphasis on agricultural pests.

532 Nota Lepidopterologica. 1977–. 4 per year. De, Fr, En. Societas Europaea Lepidopterologica, c/o G. Ebert, Landessammlungen für Naturkunde, Karlsruhe 1, Federal Republic of Germany.

All aspects of lepidopterology of the Palaearctic region. Volumes up to 1980 carried an annual bibliography of Palaearctic Lepidoptera, accompanied by a list of new taxa for the year concerned. In future this will be issued as a separate publication of the Society.

533 Notulae Entomologicae. 1921–. 4 per year. Fi, En. Societas Entomologica Helsingforsiensis, Freelsgatan 15B, SF-00170 Helsinki 17, Finland.

Original research on general entomology of Finland, book reviews, annual index.

534 Oecologia. 1924–. 16 per year. En, Fr, De. Springer-Verlag, 175 Fifth Avenue, New York, NY 10010, USA.

The journal of the International Association for Ecology. Includes a small percentage of papers on applied entomology.

535 Oriental Insects. 1968–. 4 per year. En. Association for the Study of Oriental Insects, c/o Department of Zoology, University of Delhi, Delhi 110007, India.

Original research; systematic entomology and zoogeography of south and south-east Asia, including taxonomy, ecology and evolution of insects and other land arthropods from the Oriental region.

536   Oriental Insects Monographs. 1972–. Irregular. En. Association for the Study of Oriental Insects, c/o Department of Zoology, University of Delhi, Delhi 110007, India.

To date the majority of volumes have been on Ichneumonidae. Other families are projected. Each monograph is complete in itself, and is intended to be an exhaustive study of the group under discussion.

537   Oriental Insects Supplements. 1971–. Irregular. En. Association for the Study of Oriental Insects, c/o Department of Zoology, University of Delhi, Delhi 110007, India.

Large monographic articles on specific aspects of the Oriental region's insect fauna.

538   Pacific Insects. 1959–. 4 per year. En. Department of Entomology, Bishop Museum, Honolulu, Hawaii, USA.

Original research primarily on the systematics and zoogeography of terrestrial arthropods of the Pacific area including eastern Asia, Australia and Antarctica. Annual index, occasional book reviews.

539   Pacific Insects Monographs. 1961–. Irregular. En. Department of Entomology, Bishop Museum, Honolulu, Hawaii, USA.

Numbered series of large monographs on entomology, acarology and bibliography of the subject.

540   Pakistan Journal of Agricultural Research (Agriculture Pakistan). 1980–. 4 per year. En. Pakistan Agricultural Research Council, PO Box 1031, Islamabad, Pakistan.

Original agricultural research carried out in Pakistan, including 15% economic entomology.

541   Pan-Pacific Entomologist. 1924–. 4 per year. En. Pacific Coast Entomological Society, California Academy of Sciences, Golden Gate Park, San Francisco, California 94118, USA.

Systematics and biology of entomology, mostly of North America. Book reviews and biographical material. Annual index.

542   Parasitica. 1945–. 4 per year. Fr, En. Association pour les Etudes et Recherches de Zoologie Appliquée et de Phytopathologie, Station de Phytopharmacie de l'Etat, rue du Bordia 11, B-5800 Gembloux, Belgium.

Original research and short communications on plant pathology and applied zoology, including 10% agricultural entomology.

543   Parazitologicheskii Sbornik. 1930–. Irregular. Ru, En. Library, Akademiya Nauk SSR, Birzhevaya liniya 1, 199164 Leningrad, USSR.

Original research on all aspects of parasitology including 15% medical and (predominantly) veterinary entomology.

544 Parazitologiya. 1967–. 6 per year. Ru, En. Library, Akademiya Nauk SSR, Birzhevaya liniya 1, 199164 Leningrad, USSR.

Original research, brief reports and book reviews (the latter in Russian only) on all aspects of parasitology, including 40% medical and veterinary entomology.

545 Pesticide Biochemistry and Physiology. 1971–. 6 per year. En. Academic Press, 111 Fifth Avenue, New York, NY 10003, USA.

Original research dealing with the biochemistry and physiology of pesticides, including non-lethal pest control agents such as antifeeding compounds and chemosterilants, etc. Short communications and notes are excluded. Includes 44% economic entomology. Book reviews. Annual author index, biannual (1 per 2 vols) subject index.

546 Pesticide Science. 1970–. 6 per year. En. Blackwell Scientific, Osney Mead, Oxford OX2 0EL, England.

Original research on all aspects of the production, use, metabolism, degradation and toxicology of synthetic and naturally occurring insecticides, fungicides, herbicides and growth regulations for agricultural, veterinary and public health use. Also includes papers on the ecological implications and economics of using pesticides and other methods of pest, disease and weed control. Includes lists of recently adopted BSI common names for pesticides. Annual author and subject indexes. About 15% economic entomology.

547 Pesticides. 1967–. 12 per year. En. Colour Publications Pvt, 126-A Dhurunadi, Off Dr Nariman Road, Near Bengal Chemicals, Bombay 400025, India.

Original research, brief notes and news items on all aspects of pesticide use in India. Includes some 30% economic entomology.

548 Pflanzenschutz-Nachrichten 'Bayer'. 1962–. 3 per year. En. Bayer Pflanzenschutz, Leverkusen, Federal Republic of Germany.

Reports on results of research and field trials of Bayer Agrochemicals. Annual subject index. Includes 20% economic entomology (mainly agricultural).

549 Phegea. 1973–. Irregular. Nd, Fr, En. W. de Prins, Diksmuidelaan 176, B-2600 Berchem, Belgium.

General entomology, with emphasis on Lepidoptera; short notes on collecting and distribution of the local fauna.

550 Philippine Entomologist. 1968–. Irregular. En. Philippine Association of Entomologists, c/o Department of Entomology, University of the Philippines, College Lugana, Philippines 3720.

Six parts per volume, but each volume is published over several years and publication dates are late. Papers are based on original research on all aspects of the subject; about 5% are on arachnids. Strong emphasis on economic entomology.

551   Physiological Entomology. 1976–. 4 per year. En. Blackwell Scientific, for the Royal Entomological Society of London, 41 Queen's Gate, London SW7 5HU, England.

Original research, experimental analysis of behaviour, general physiology, endocrinology, behavioural physiology and biochemistry.

552   Phytopathology. 1911–. 12 per year. En. American Phytopathological Society, 3340 Pilot Knob Road, St Paul, Minnesota 55121, USA.

An international journal of phytopathology. Includes papers on insect pests. Book reviews.

553   Plant Pathology. 1952–. 4 per year. En. Blackwell Scientific, Osney Mead, Oxford OX2 0EL, England.

Original UK-based research on, and new and unusual records of, plant pests and diseases in the British Isles. Includes 30% agricultural entomology, separate annual author and subject indexes.

554   Plant Protection Bulletin, FAO. 1952–. 4 per year. En, Fr, Es (separate editions). Ford and Agriculture Organization, via delle Terme di Caracalla, I-00100 Roma, Italy.

Reports on the occurrence, outbreak and control of pests of economic significance, plant quarantine announcements and related topics, with special reference to current information. Includes 55% economic entomology.

555   Plant Protection Bulletin, Taiwan. 1959–. 4 per year. En, Ch. Plant Protection Society of the Republic of China, c/o Department of Entomology and Plant Pathology, National Taiwan University, Taipei, Taiwan, Republic of China.

Original Chinese research on plant pathology and related fields. Includes 35% agricultural entomology.

556   Polskie Pismo Entomologiczne. 1922–. 4 per year. Pl, De, En. Dr W. Pulawski, ul. Sienkiewicza 21, 50-335 Wrocław, Poland.

Original research, insect fauna of the Palaearctic region, general and applied entomology.

557   Probleme de Protectia Plantelor. 1973–. 4 per year. Ro, En. Oficiul de Informare Documentara Pentru Agricultura si Industire Alimentara, Bd. Marasti 61, Bucureşti 1, Romania.

Original research and review articles on Romanian plant protection including 50% agricultural entomology.

558   Proceedings, Annual Meetings of the Utah Mosquito Abatement Association. 1948–. 1 per year. En. Utah Mosquito Abatement Association, PO Box 983, Vernal, Utah 84048, USA.

Original research papers presented at the Association's annual meetings, together with Association business matters. 100% medical/veterinary entomology.

559 Proceedings of the Association for Plant Protection of Kyushu. 1953–. 1 per year. Ja, En. Association for Plant Protection of Kyushu, c/o Kyushu Agricultural Experiment Station, Chikugo, Fukuoka Prefecture, 833, Japan.

Short papers reporting original research on all aspects of plant protection in Japan. Includes 40% agricultural entomology.

560 Proceedings, California Mosquito and Vector Control Association. 1933–. 1 per year. En. California Mosquito and Vector Control Association, 1737 West Houston Avenue, Visalia, California 93211, USA.

Original research in the field of mosquito and related vector control, papers and presentation of the Association's annual conference and reports on the Association's activities. 100% medical/veterinary entomology.

561 Proceedings of the Entomological Society of Washington. 1884–. 4 per year. En. Entomological Society of Washington, Department of Entomology, Smithsonian Institution, Washington, DC 20560, USA.

Original research, general entomology including ticks, a strong emphasis on taxonomy. Book reviews.

562 Proceedings of the Hawaiian Entomological Society. 1906–. Irregular. En. Hawaiian Entomological Society, Department of Entomology, University of Hawaii, 2500 Dole Street, Honolulu, Hawaii 96822, USA.

Original research on fields of entomological research, emphasis on Hawaiian fauna, some papers on economic entomology.

563 Proceedings of the Royal Entomological Society. 1847–1977 (1847–1932, published as *Proceedings of the Entomological Society*). 1936–70 issued in series: Series A, *General Entomology*; Series B, *Taxonomy*; Series C, *Proceedings of Meetings*. 1971–77, *Journal of Entomology, Series A* and *B*.

This serial has now ceased publication, but its contents include many important papers.

564 Proceedings and Transactions of the British Entomological and Natural History Society. 1968–. 4 per year. En. The Editor, British Entomological and Natural History Society, c/o The Alpine Club, 74 South Audley Street, London W1Y 5FF, England.

Papers are 95% entomological. Original research and reports of meetings including field meetings. Book reviews and obituaries of well-known members. Annual index.

565 Proceedings and Transactions of the South London Entomological and Natural History Society. 1885–1967. En. Predates 564.

Original research on British natural history. Some 90% entomology. Reports of meetings and field meetings. Annual index.

566 Protection Ecology. 1978–. 4 per year. En. Associated Scientific Publishers, PO Box 211, Amsterdam, Netherlands.

Original research, review articles, short communications, editorials and book reviews. Aims to cover all aspects of the study and management of noxious organisms in plant and animal industries. Includes 45% economic entomology.

567 Psyche. 1874–. 4 per year. En. Cambridge Entomological Club, Biological Laboratories, Harvard University, Cambridge, Massachusetts 02138, USA.

Papers on original research, some emphasis on the Neotropical fauna. Covers all orders of insects plus papers on mites and spiders. Book reviews.

568 Quaestiones Entomologicae. 1965–. 4 per year. En, Fr. G.E. Ball, Department of Entomology, University of Alberta, Edmonton, Alberta T6G 2E3, Canada.

Original research, taxonomy of all groups of insects. Not restricted to the Canadian fauna. Book reviews. Annual index.

569 Research Bulletin of the Punjab University, Zoology. 1954–. 2 per year. En. Punjab University, Chandigarh 160014, India.

Publishes original research papers, short notes and reviews by students and staff of the Punjab University. Includes 20% economic entomology (mostly agricultural).

570 Researches on Population Ecology. 1952–. Irregular. En, Ja. Society of Population Ecology, c/o Japan Academic Societies Center, Yayoi 2-4-16, Bunkyo-ku, Tokyo 113, Japan.

Original research on all aspects of plant and animal population ecology, including 80% economic entomology.

571 Revista Brasileira de Entomologia. 1954–. 2 per year. Pt, Es, En. Sociedade Brasileira de Entomologia, Caixa Postal 9063, 01.1000 São Paulo, SP, Brazil.

Original research, pure and applied taxonomy, including associated groups of arthropods, restricted to the Brazilian fauna.

572 Revista Brasileira de Malariologia e Doencas Tropicais. 1949–. 4 per year (sometimes issued annually). Pt, Es. Servico de Divulgacao e Informacao, Avenida Brasil 4036, 4° andar, Rio de Janeiro, Brazil.

Original research, control of insect vectors. Emphasis on epidemiology of malaria.

573 Revista de la Sociedad Entomológica Argentina. 1926–. 1 per year. Es, En. Sociedad Entomológica Argentina, Casolla de Correo No. 1, Sucursal No. 2, Buenos Aires, Argentina.

Original research on systematic entomology, mostly on South American fauna. Includes papers on Acarina.

574   Revista de la Sociedad Mexicana de Lepidopterología. 1975–. 4 per year. Es, En. Sociedad Mexicana de Lepidopterología, Nicolas San Juan 1707, Mexico.

Original scientific contributions relating to American Lepidoptera. Short notes and book reviews.

575   Revue Agricole et Sucrière de l'Ile Maurice. 1955–. 4 per year. Fr, En. La Revue Agricole et Sucrière de l'Ile Maurice, c/o MSIRI, Reduit, Mauritius.

Original research on agriculture (especially sugar cane) in Mauritius, including 18% economic entomology. Advertising.

576   Revue d'Elevage et de Médecine Vétérinaire des Pays Tropicaux. 1947–. 4 per year. Fr, En, Es. Editions Vigot, 23 rue de l'Ecole de Médecine, F-75006 Paris, France.

Original research and abstracts. Includes 15% medical and veterinary entomology.

577   Revue Française d'Entomologie. 1934–65; 1979–. 4 per year. Fr. Association des Amis du Laboratoire d'Entomologie du Muséum, 45 rue de Buffon, F-75005 Paris, France.

Papers on the research of the staff on the collections of the Laboratoire d'Entomologie, Muséum Nationale d'Histoire Naturelle.

578   Revue de Zoologie Agricole et de Pathologie Végétale. 1969–. 4 per year. Fr, En. Société de Zoologie Agricole, Domaine de la Grande Ferrade, F-33140 Pont de la Maye, France.

Original French research on agricultural zoology and phytopathology, including 90% agricultural entomology.

579   Rivista di Parassitologia. 1937–. Irregular. En, It, Fr, Es, De. Istituto di Parassitologia, via Cesare Battisti 48, I-98100 Messina, Italy.

Original research, reviews, research notes, news items and book reviews on protozoology, helminthology, entomology, mycology and parasitic diseases. Includes 10% economic entomology.

580   Southeast Asian Journal of Tropical Medicine and Public Health. 1970–. 4 per year. En. SEAMEO Regional Tropical Medicine and Public Health Project, 4206 Rajivithi Road, Bangkok 4, Thailand.

Original research, brief communications, correspondence and news on parasitology, tropical medicine and public health in south-east Asia. Includes 20% medical entomology.

581   Southwestern Entomologist. 1976–. 4 per year. En. Southwestern Entomological Society, Department of Entomology, Texas A & M University, College Station, Texas 77840, USA.

Original research relating to entomology (largely economic) in the southwestern United States and Mexico. Annual author and subject indexes.

582   Special Bulletin of the Lepidopterological Society of Japan. 1965–.
Irregular. En. Lepidopterological Society of Japan, Osaka, Japan.

Large monograph-style papers. Reports of expeditions and taxonomic revisions. Papers restricted to Japanese Lepidoptera.

583   Special Publication of the Japanese Hymenopterists' Association.
1976–. Irregular. En. K. Tsuneki, Asahigaoka 4–15, Mishima, Japan
411.

Numbered monographs. Papers restricted to the Japanese fauna.

584   Systematic Entomology. 1976–. 4 per year. En. Blackwell Scientific,
for the Royal Entomological Society of London, 41 Queen's Gate,
London SW7 5HU, England.

Papers on taxonomy and systematics with an emphasis on comprehensive and revisionary studies and work with a biological and zoogeographical relevance.

585   Systematic Zoology. 1952–. 4 per year. En. Society of Systematic
Zoology, c/o Smithsonian Institution, 10th & Constitution NW,
Washington, DC 20560, USA.

Advancement of systematic zoology in all its aspects of theory, principles, methodology and practice for both living and fossil animals with emphasis on areas of common interest to all taxonomists regardless of individual specialization. An important journal for articles of a review nature. Carries excellent book reviews.

586   Technical Bulletin, United States Department of Agriculture. 1927–.
Irregular. En. Science and Education Administration, Washington,
DC 20250, USA.

Each issue is devoted to a single, often substantial, original research paper or, occasionally, a review or bibliography, by a USDA author. Approximately 15% economic entomology.

587   Tijdschrift voor Entomologie. 1857–. Irregular. En, Nd. Nederlandse
Entomologische Vereniging, Rijksmuseum van Natuurlijke Historie,
Raamsteeg 2, NL-2311 PL Leiden, Netherlands.

Publishes papers on Insecta, Myriapoda and Arachnoidea. Original research not restricted to any zoogeographic area.

588   Tinea. 1953– 2 per year. Ja, En. Japan Heterocerists' Society, c/o
National Science Museum, Ueno Park, Tokyo, Japan.

Papers on taxonomy of moths. The distribution, biology, behaviour and migration of Japanese Heterocera.

589   Transactions of the American Entomological Society. 1867–. 4 per
year. En. American Entomological Society, at the Academy of Natural
Sciences, Philadelphia, Pennsylvania 19103, USA.

Original research. Each quarterly part usually contains two to three large papers in systematic research.

Transactions of the Lepidopterological Society of Japan – *see* **Tyo to Ga** (entry 602)

590   Transactions of the Royal Entomological Society of London. 1807–1812, 1834–1975 (1834–1933, published as *Transactions of the Entomological Society*).

This journal has now ceased publication but through its long series are many important papers on taxonomic entomology and the biology of insects.

591   Transactions of the Royal Society of Tropical Medicine and Hygiene. 1907–. 6 per year. En. Royal Society of Tropical Medicine and Hygiene, Manson House, 26 Portland Place, London W1N 4EY, England.

Original research, brief communications, correspondence and reports of meetings on health and diseases in developing countries, their health services, tropical medicine and hygiene, the communicable diseases of the tropics and their vectors, and those aspects of human and animal biology which relate to tropical diseases, with special emphasis on their epidemiology and control and on clinical and experimental work. Includes 12% medical/veterinary entomology.

592   Transactions of the Shikoku Entomological Society. 1950–. Irregular. En, Ja. Shikoku Entomological Society, Entomological Laboratory, College of Agriculture, Ehime University, Matsuyama, Japan.

Original research, mostly taxonomic entomology of Japan and neighbouring countries and islands.

593   Treballs de la Societat Catalana de Lepidopterologia. 1980–. 4 per year. Ca, Es. A. Maso i Planas, Societat Catalana de Lepidopterologia, Copisteria Castella, Pujol 39, Mataro, Spain.

Original research on Lepidoptera of the Catalan area.

594   Tropenmedizin und Parasitologie. 1974–. 4 per year. De, En. Georg Thieme Verlag, Herdweg 63, Postfach 732, D-7000 Stuttgart 1, Federal Republic of Germany.

The journal of the Deutsche Tropenmedizinische Gesellschaft. Original research and book reviews on tropical medicine and parasitology, including 25% medical/veterinary entomology.

595   Tropical Grain Legume Bulletin. 1975–. 4 per year. En. International Grain Legume Information Centre, IITA, PMB 5320, Ibadan, Nigeria.

Published by the International Institute of Tropical Agriculture and designed as a forum for research and extension workers. Includes original research, short notes, news items and abstracts of current literature on: cowpeas, lima beans, pigeon peas, African yam bean, groundnuts, Mexican yam bean, Baubara groundnut, winged bean, Kersting's groundnut, locust bean, and soybean cultivation in the tropics only. Approximately 20% agricultural entomology.

596 Tropical Medicine. 1967–. 4 per year. En, Ja. Institute for Tropical Medicine, Nagasaki University, 12-4 Sakamoto-machi, Nagasaki 852, Japan.

Original research carried out at the Institute for Tropical Medicine, Nagasaki University. 25% medical/veterinary entomology.

597 Tropical Pest Management (formerly **PANS/Pest Articles and News Summaries**). 1980–. 4 per year. En. Centre for Overseas Pest Research, College House, Wrights Lane, London W8 5SJ, England.

Original research, reviews, short communications, notes and news. Covers the entire field of pest management in its broadest sense, including the biology and control of weeds, plant diseases, nematodes and vertebrate pests in agriculture, and insects in agriculture and public health, with particular reference to the tropics and subtropics. Useful sections include: conference reports, new books, and pesticide application equipment information sheets. Includes a limited amount of trade advertising. Annual index of authors, subjects, conferences, and new publications noticed.

598 Tropical Stored Products Information. 1960–. Irregular. En. Tropical Products Institute, Tropical Stored Products Centre, London Road, Slough SL3 7HL, England.

Original research, reviews, bibliographies, abstracts and book reviews. Covers the storage of durable agricultural produce in tropical countries, including stored-product entomology.

599 Trudy Vsesoyuznogo Entomologicheskogo Obschestva. 1951–. Irregular. Ru. Horae Societatis Entomologicae Unionis Soveticae, Academy of Sciences of the USSR, Leningrad, USSR.

Each volume usually contains a collection of papers written around a collective theme or title. All papers in Russian and all on the fauna of the USSR.

600 Tsetse and Trypanosomiasis Information Quarterly. 1978–. 4 per year. En, Fr (separate editions). Centre for Overseas Pest Research, College House, Wrights Lane, London W8 5SJ, England.

Includes news items, abstracts and bibliographic references relating to all aspects of tsetse and trypanosomiasis control. Separate annual subject and author indexes.

601 Turkiye Bitki Koruma Dergisi (Turkish Journal of Plant Protection). 1977–. 4 per year. Tu, En. E.U. Ziraat Fakultesi Entomologi, ve Zirai Zoologi, Kursusu, Izmir, Turkey.

Economic and taxonomic entomology of the Turkish or allied fauna.

602 Tyo to Ga (Butterflies and Moths. Transactions of the Lepidopterological Society of Japan). 1945–. 4 per year. En, Ja. Lepidopterological Society of Japan, c/o Ogata Hospital, 18 Imabashi 3 chamer, Higashi-ku, Osaka 541, Japan.

Taxonomy and distribution of butterflies and moths of Japan and adjacent territories. Although nominally this is a quarterly journal, parts are often issued on an annual basis.

603   University of California Publications. Entomology. 1906–. Irregular. En. University of California Press, 2223 Fulton Street, Berkeley, California 94720, USA.

Numbered and separately issued. Authoritative monographs.

604   Zeitschrift für Angewandte Entomologie. 1914–44; 1949–. 12 per year. En, Fr, De. Verlag Paul Parey, Spitalerstr. 12, D-2000 Hamburg 1, Federal Republic of Germany.

Covers all aspects of applied entomology, with emphasis on biological control. Includes book reviews.

605   Zeitschrift der Arbeitsgemeinschaft Österreichischer Entomologen (incl. **Entomologisches Nachrichtenblatt**). 1960–. 2 per year. De, En, Fr. Arbeitsgemeinschaft Österreichischer Entomologen, Ludo-Hartmann-Platz 7, A-1160 Wien, Austria.

Original research, mostly Palaearctic fauna. Largely taxonomic entomology.

606   Zimbabwe Journal of Agricultural Research (Rhodesian Journal of Agricultural Research). 1963–. 2 per year. En. R. & S.S. Information Services, Department of Research and Specialist Services, Ministry of Agriculture, PO Box 8108, Causeway, Harare, Zimbabwe.

Original research, research notes and reviews in the fields of soil and renewable natural resources. Includes 15% economic entomology. Book reviews.

BOOKS AND REVIEW ARTICLES

The selection of titles for inclusion in this section has perhaps been the most difficult for the whole book. So many titles appear each year, and it is impossible in a volume of this size to include them all. Thus a small selection is offered; we have tried to cover texts in languages other than English and to give some titles which are aimed at the amateur or beginner as well as the student and the research worker.

For the convenience of the user, we have broken this section into a number of subheadings by subject. However, it should be remembered that many of these works can usefully be used in conjunction with each other; perhaps a more general textbook alongside a specialized volume.

Many of these works are well referenced, and judicious use of their bibliographies will lead the user to further reading. Use of some of the entomological secondary services – abstracting journals or on-line retrieval services – will lead the user to relevant articles in scientific journals.

Fundamental entomology

GENERAL

607 Atkins, M.D. 1978. *Insects in perspective.* vii + 513pp. Macmillan; New York.

Integrated coverage of pure and applied entomology. A fairly good treatment of behaviour and ecology, and the beneficial and harmful effects on man. The work is generously illustrated with line drawings, but many have not reproduced well. There is a general and a taxonomic index. On pp. 488–500 is a short glossary of terms used in the work.

Blackwelder, R.E. *Taxonomy, a text and reference book – see* entry 45

608 Borror, D.J., DeLong, D.M. and **Triplehorn, C.A.** 1981. *An introduction to the study of insects.* 5th edn. xi + 827pp. W.B. Saunders; Philadelphia.

A good entomological textbook intended as a student's text for the American college. Enough general entomology, i.e. morphology, physiology, etc., is presented to enable students to use keys for the identification of families of insects occurring in North America north of Mexico. This 5th edition has been thoroughly revised and updated in many sections. The bibliography is not overgenerous, but is intended only as a starting point for further reading.

Chapman, R.F. *The insects: structure and function – see* entry 739

609 Corbet, P.S. 1962. *A biology of dragonflies.* 247pp. H.F. & G. Witherby; London.

The volume is mainly concerned with the ecology of dragonflies and in particular their physiology and behaviour. The work is a good general text aimed mainly at the undergraduate. It has a good bibliography.

610 Crawford, C.S. 1981. *Biology of desert invertebrates.* 314pp. Springer-Verlag; Berlin.

The volume covers groups other than insects in equal fashion. A relatively small work to deal with this large and expanding area of interest, but the biology and ecology are covered fairly succinctly. Includes a good bibliography.

611 Crowson, R.A. 1981. *The biology of the Coleoptera.* xii + 802pp. Academic Press; London.

An up-to-date summary of biological knowledge of the Coleoptera. An extensive bibliography up to the beginning of 1980 is included. A work of reference for the Coleoptera specialist.

612 CSIRO [Corporate authorship]. 1970. *The insects of Australia: a textbook for students and research workers.* xiii + 1029pp. Supplement 1974. viii + 146pp. Melbourne University Press; Melbourne.

The work uses the Australian fauna, but it can be used as a text on a world-wide basis. It is a fine textbook, with good illustrations. There is a very full bibliography, combined at the end of the volume.

613 Daly, H.V., Doyen, J.T. and **Ehrlich, P.R.** 1978. *Introduction to insect biology and diversity.* 546pp. McGraw-Hill; New York.

A good text on all subject groups. The work is mainly aimed at the undergraduate in American state colleges. The line drawings are numerous and good. There is a broad-based bibliography; a subject and taxonomic index is included.

614 Eisner, T. and **Wilson, E.O.** (Eds). 1977. *The insects.* 334pp. W.H. Freeman; San Francisco. (*Readings from 'Scientific American'.*)

A collection of review articles written originally for *Scientific American.* Brought together they form a nice broad review of most aspects of entomology. The articles are of a semi-popular nature and make a good introduction to the subject. In the main, they are written by well-known, authoritative authors, for example Wigglesworth on metamorphosis and differentiation, Bruce-Chwatt on malaria and Rothschild on fleas. The illustrations are very good.

615 Elzinger, R.J. 1981. *Fundamentals of entomology.* 2nd edn. x + 422pp. Prentice-Hall; Englewood Cliffs, New Jersey.

Written as a student text for the American colleges, it is a good introduction to entomology. Each of the chapters has questions posed at the end. The work includes keys to the families of insects and a chapter on making an insect collection. On pp. 393–400 is a glossary of terms used in the text.

616 Feltwell, J. 1981. *The large white butterfly: the biology, biochemistry and physiology of* Pieris brassicae (*L.*). 542pp. Dr W. Junk; The Hague. (*Series Entomologica* No. **18.**)

'A source book for relevant literature and an introduction to all aspects of the white butterfly.' Some 41 000 references to key papers on the insect are included.

617 Gillott, C. 1980. *Entomology.* 729pp. Plenum Press; New York.

This work covers evolution and diversity, anatomy and physiology, reproduction, development and ecology. The coverage tends to be rather unbalanced. An undergraduate text.

618 Grandi, G. 1951. *Introduzione allo studio dell'entomologia.* 2 volumes. Vol. 1, 950pp. Vol. 2, 1332pp. Edizioni Agricola; Bologna. Additions and corrections are to be found in *Bollettino dell'Istituto di Entomologia della Università di Bologna* **19** (1953): 263–306.

An Italian textbook for undergraduate and research worker alike. The work is well illustrated; the photographs, although fewer in number than the line diagrams, have not reproduced well. There is a good bibliography and index.

619   Grandi, G. 1966. *Istituzioni di entomologia generale.* 655pp. Edizioni Caeldaerini; Bologna.

An Italian text. Most of the text and illustrations suggest it to be a précis version of this author's 1951 *magnum opus* (entry 618). There is no bibliography.

620   Grassé, P.P. 1948–. *Traité de zoologie. Insecta.* Masson; Paris.

This large, comprehensive reference work on zoology must surely be the most detailed of its kind. It covers all aspects of the subject. Volumes are issued at irregular intervals; the Insecta volumes among others are still being issued. The work is well illustrated and carries good bibliographies. It is useful both to undergraduate and research worker.

621   Herreid, C.F., II and **Fourtner, C.R.** (Eds). 1981. *Locomotion and energetics in arthropods.* 546pp. Plenum; New York.

Basically the proceedings of a symposium held in 1980, described by the editors as one of the first attempts to deal with all of the major modes of arthropod locomotion in a comprehensive fashion.

622   Imms, A.D. 1971. *Insect natural history.* 3rd edn. 317pp. Collins; London. (Collins' New Naturalist series.)

A good general, semi-popular account of insects: how they are classified; their general biology; social insects and biological control. An eminently useful text for the lay reader or beginner. It is well illustrated.

623   Imms, A.D. 1977. *A general textbook of entomology: including the anatomy, physiology, development and classification of insects.* 10th edn, revised by O.W. Richards and R.G. Davies. 2 volumes. Chapman & Hall; London.

This well-known textbook for both undergraduates and research workers is now published in two volumes. Some parts are barely revised for the 10th edition, and the bibliographies have not been well up-dated. The work does not contain keys, which are a feature of many entomological textbooks.

624   Kükenthal, W. (Ed.). 1969–. *Handbuch der Zoologie*; Band IV, *Arthropoda* – 2. Hälfte, *Insecta.* Walter de Gruyter; Berlin. New edition edited by J.-G. Helmcke, D. Starck and H. Wermuth.

The work is issued at irregular intervals in parts. There are separate authors for individual families. This German textbook has a very thorough coverage of the subject. It is well illustrated, with good up-to-date bibliographies.

625   Lara, F.M. 1979. *Principios de entomologia.* [In Portuguese]. 295 + [9]pp. Livoceres; Piracicala.

The format is very clear, but the illustrations are poor, particularly the photographs. There is a slight bias towards economic entomology; the work has a reasonably up-to-date bibliography.

626   McCafferty, N.P. 1981. *Aquatic entomology: the fishermen's and* *ecologists' illustrated guide to insects and their relatives.* 448pp. Science Books International; Boston, Massachusetts.

Written in a semi-popular, easy-to-read style. The work is illustrated and as its title suggests, much emphasis is given to insects associated with fishing. The coverage is fairly detailed, with much information offered on biology. Although it has many illustrations, on the whole they are rather poor. Each chapter has a good bibliography. There is heavy use of American common names. A glossary is included and an appendix of common and scientific names.

627 Oldroyd, H. 1964. *The natural history of flies.* 324pp. Weidenfeld & Nicolson; London.

A most readable introductory text to this group of insects. The work covers general life histories and all groups of the flies. A useful bibliography is included.

628 Romoser, W.S. 1981. *The science of entomology.* 2nd edn. 575pp. Macmillan; New York and London.

A useful introductory text, aimed at the American college undergraduate. It is of general use, well illustrated, with the diagrams very clearly labelled. Well indexed.

629 Rozkosny, R. (Ed.). 1980. *Keys to water larvae of insects.* [In Czech]. 518pp. Československá Akademie Věd; Prague.

The language may be a barrier to some users, but the coverage of insect groups is extremely good. The work is extensively illustrated, but the quality of reproduction is poor.

630 Séguy, E. 1950. *La biologie des diptères.* 609pp. P. Lechevalier; Paris. (*Encyclopédie entomologique* XXVI.)

A superb volume of reference for the biology of the Diptera. All families are covered. The work includes some excellent colour plates. The numbers in the very full index refer to numbered paragraphs in the main body of the work and not to page numbers.

Skaife, S.H. *African insect life – see* entry 89

631 Tweedie, M. 1977. *Insect life.* 192pp. Collins; London.

One of a series, aimed at the beginner in natural history. A good basic text for the amateur, but not written in a patronizing fashion. There is a good list of references for further reading, most of them up to date and easy to acquire.

632 Villiers, A. 1979. *Initiation à l'entomologie.* 2 volumes. Société Nouvelle des Éditions Boubée; Paris.

A semi-popular text in French aimed mainly at the amateur. It is a good introduction to the subject. The black-and-white illustrations are poor, but the colour good.

633   Wigglesworth, V.B. 1964. *The life of insects.* xii + 360pp. Weidenfeld & Nicolson; London.

A popular text on insects and most useful to the beginner or amateur. The work does not purport to be a textbook, but it is a very clear exposition of the subject. It is well illustrated and the bibliography for each chapter is annotated for further reading interests.

ACOUSTICS – INSECT SOUNDS

634   Alexander, R.D. 1967. Acoustical communication in Arthropods. *Annual Review of Entomology* **12**: 495–526.

A highly selective review article with a good bibliography. On pp. 500–504 is a glossary of terms for describing arthropod acoustical signals.

635   Busnel, R.-G. 1955. *Colloque sur l'acoustique des orthoptères.* 448pp. Institut National de la Recherche Agronomique; Paris.

Much of the acoustic work has been done with Orthoptera, but this work is also useful on a more general level. It has a good bibliography, but is not indexed.

636   Busnel, R.-G. (Ed.). 1963. *Acoustic behaviour in animals.* 933pp. Elsevier; Amsterdam.

The work has a very good subject index, using both the vernacular and specific names; there is also a 'glossarial index'. Terms are not explained in the index, but reference is made back to the appropriate chapter in the book. A systematic index is also included.

637   Frings, M. and **Frings, H.** 1960. *Sound production and sound reception by insects: a bibliography.* 108pp. Pennsylvania State University Press; University Park, Pennsylvania.

An alphabetical index by author, but also cross-referenced with a taxonomic index. The authors claim the index to be complete to 1957. A considerable number of papers have been published since that time, but it is a good start for the early literature.

638   Haskell, P.T. 1961. *Insect sounds.* 189pp. H.F. Witherby; London.

Written primarily as a basic student text. A most useful introduction to the subject.

639   Lanyon, W. E. and **Tavolga, W. N.** 1960. *Animal sounds and communication.* 443pp. American Institute of Biological Sciences; Washington, DC. (Publication No. 7.)

This work contains a good chapter on recording sound for bioacoustics studies and sound communication in Orthoptera and Cicadidae.

640   Pierce, G.W. 1948. *The sounds of insects; with related material on the production, propagation, detection and measurement of supersonic vibration.* vii + 329pp. Harvard University Press; Cambridge, Massachusetts.

One of the earliest books on the subject, but some of the work described in this volume has not been repeated in later books.

641 Sebeok, T.A. 1977. *How animals communicate.* 1128pp. Indiana University Press; Bloomington, Indiana.

The early chapters of the work deal with animal communication in general terms, Chapters 16–19 with selected orders of insects. Each chapter has a good bibliography.

642 Snodgrass, R.E. 1923. Insect musicians, their music and their instruments. *Annual Report of the Smithsonian Institution* **1923**: 405–52.

A semi-popular account of insect songs; a useful introduction to the subject and not too technical.

643 Tuxen, L. 1967. *Insektenstimmen.* 156pp. Springer-Verlag; Berlin.

A good introductory work. A rather poor bibliography.

644 Zhatiev, R.D. 1981. *Bioacoustics of insects.* [In Russian]. 256pp. Moscow University Press; Moscow.

The volume is divided into three main parts: acoustic communication systems in insects, recognition of acoustic signals, and sound source location. The final chapter discusses the use of sound in pest control. The work has a good, full bibliography.

BEHAVIOUR, INCLUDING SOCIAL BIOLOGY

645 Askew, R.R. 1971. *Parasitic insects.* xvii + 316pp. Heinemann; London.

This volume, the only one to cover parasitic insects as a whole, is a good general account aimed at the university undergraduate and postgraduate student. The quality of the illustrations varies considerably. There is a good up-to-date bibliography and the work is well indexed.

646 Blum, M.S. and **Blum, N.A.** (Eds). 1979. *Sexual selection and reproductive competition in insects.* 463pp. Academic Press; New York.

A useful collection of papers derived from a symposium held at the 15th International Congress of Entomology.

647 Denno, R.F. and **Dingle, H.** (Eds). 1981. *Insect life history patterns: habitat and geographic variation.* 225pp. Springer-Verlag; New York.

This volume 'offers a thorough understanding of the evolutionary problems environmental heterogeneity poses for insects and the complexity and shortcomings of life history theory'.

648 Edmunds, M. 1974. *Defence in animals: a survey of anti-predator defences.* 357pp. Longman; London.

Describes the ingenious means whereby animals avoid being eaten, many examples being taken from the Insecta. The complex interactions between predators and prey are explained. Well illustrated.

649   Free, J.B. 1970. *Insect pollination of crops.* 544pp. Academic Press; London.

650   Haskell, P.T. (Ed.). 1966. Insect behaviour. *Symposia of the Royal Entomological Society of London* **3**: 1–113.

Eight papers by specialists in their own field, including flight behaviour, feeding and sexual behaviour, communication, and communication in social insects.

651   Hermann, H.R. (Ed.). 1978–82. *Social insects.* 4 volumes. Academic Press; New York.

When completed, this should be an important and fairly exhaustive work on sociobiology.

652   Howse, P.E. and **Clement, J.L.** 1981. Biosystematics of social insects. *Systematics Association Special Volume* No. **19**: 1–346.

A collection of 26 papers given at a symposium, giving a good coverage of the subject. Each paper has its own full bibliography.

653   Kettlewell, H.B.D. 1973. *The evolution of melanism: the study of recurring necessity, with special reference to industrial melanism in the Lepidoptera.* xxiv + 423pp. Clarendon Press; Oxford.

Standard basic text on the subject, written by the foremost authority on melanism. The text is not too specialized and can be used by student and specialist alike.

654   Matthews, R.W. and **Matthews, J.R.** 1978. *Insect behaviour.* 507pp. John Wiley; New York.

A good comprehensive text, but the illustrations are rather poor. Each chapter has a selective bibliography for further reading. Aimed at the undergraduate rather than the research worker.

655   Mills, J.N. (Ed.). 1973. *Biological aspects of circadian rhythms.* xiv + 319pp. Plenum; London.

Nine authors review separate aspects within their own fields of study. Chapter 6 by J.E. Harker deals specifically with circadian rhythms in insects. Chapter 1 deals with laboratory techniques and rhythmometry.

656   Mitchell, R. 1981. Insect behaviour, resource exploitation and fitness. *Annual Review of Entomology* **26**: 373–96.

A brief review of the subject. A useful bibliography.

657   Muirhead-Thomson, E.C. 1982. *Behaviour patterns of blood-sucking flies.* 240pp. Pergamon; Oxford.

Includes chapters on mosquitoes, tsetse flies, horseflies and deerflies, blackflies and phlebotomine sandflies, and a bibliography of almost 300 references.

658   Richards, A.J. (Ed.). 1978. *The pollination of flowers by insects.* 213pp. Academic Press; London. (Linnean Society Symposium Series No. 6.)

An important collection of papers on a far-ranging subject. The subject cover includes insect behaviour, flower colour polymorphism and the role of visiting insects.

659 Saunders, D.S. 1982. *Insect clocks.* 2nd edn. 409pp. Pergamon Press; Oxford.

A good source book on photoperiodism and has been well updated from the first edition. There is a brief but useful glossary, and a good appendix listing insects exhibiting rhythmic activity or photoperiodic control. An excellent bibliography is provided.

660 Schmidt, G.H. 1974. *Sozialpolymorphismus bei Insekten: Problem der Kastenbildung im Tierreich.* 974pp. Wissenschaftliche Verlagsgesellschaft; Stuttgart.

Full and comprehensive treatise on the subject, but the bibliographies are not very full. The work has a good index.

661 Shorey, H.H. and **McKelvey, J.J.** (Eds). 1977. *Chemical control of insect behaviour: theory and application.* 414pp. John Wiley; New York.

'State of the art' collection of papers, mostly on pheromones, behaviour responses to chemicals and the evaluation of how behaviour-modifying chemicals can be used in pest management systems.

662 Williams, C.B. 1958. *Insect migration.* 235pp. Collins; London.

A most readable and authoritative work by the leading expert of his time on insect migration. Useful to students and specialist alike.

663 Wilson, E.O. 1971. *The insect societies.* 548pp. Belknap Press of Harvard University Press; Cambridge, Massachusetts.

A successful attempt to synthesize insect sociology. Twenty-two chapters cover in some detail the families of social insects; there are also chapters on communication among societies, the genetic theory of social behaviour and symbiosis among social insects. There is a very full bibliography, but it is all at the end of the volume. There is some cross-referencing from the main body of the text, but it might have been more useful had the bibliography been broken down and added to each chapter. On pp. 461–71 is a glossary of terms used in insect sociology and general entomology.

CYTOGENETICS (*see also* Genetics)

664 Animal cytogenetics; Vol. 3, *Insecta*:
    Pt 1. **G.M. Hewitt.** 1979. *Orthoptera: grasshoppers and crickets.* 170pp.
    Pt 2. **M. White.** 1976. *Blattodea, Mantodea, Isoptera, Grylloblattodea, Phasmatodea, Dermaptera and Embioptera.*
    Pt 5. **S.G. Smith** and **N. Virkki.** 1978. *Coleoptera.* 366pp.
    Pt 6. **N. Ueshima.** 1979. *Hemiptera 3: Heteroptera.* 117pp.

Pt 7. **R.H. Crozier.** 1975. *Hymenoptera.* 95pp. Gebrüder Borntraeger; Berlin.

A most useful series issued at irregular intervals.

665   Blackman, R.L., Hewitt, G.M. and **Ashburner, M.** (Eds). 1980. Insect cytogenetics. *Symposia of the Royal Entomological Society of London* **10**: 1–278.

A review of the recent advances in insect cytogenetics from all over the world. The work is useful to specialist and non-specialist alike. It refers especially to grasshoppers, aphids, coccids, fruit flies, sandflies, tsetse flies and mosquitoes.

666   Darlington, C.D. and **La Cour, L.F.** 1970. *The handling of chromosomes.* 6th edn. 180pp. Allen & Unwin; London.

A useful work for information on handling chromosomes and the necessary techniques.

667   Lewis, K. R. and **John, B.** 1972. *The matter of Mendelian heredity.* 2nd edn. Longmans; London.

Clearly written elementary text, extremely well illustrated.

668   White, M.J.D. 1973. *Animal cytology and evolution.* 3rd edn. Cambridge University Press; Cambridge.

The only comprehensive treatise on animal cytogenetics summarizing and discussing knowledge of the cytogenetics of all groups of insects. Excellent bibliography. A most important work on the subject.

ECOLOGY

669   Cheng, L. (Ed.). *Marine insects.* xii + 581pp. North-Holland/Elsevier; Amsterdam.

Other intertidal air-breathing arthropods are also considered. The first six chapters are on topics of general interest, particularly ecological aspects. The second deals with the major groups of marine insects. Each chapter has its own bibliography. There is a subject and taxonomic index.

670   Emden, H.F. Van (Ed.). 1972. Insect/plant relationships. *Symposia of the Royal Entomological Society of London* **6**: 1–215.

This volume, not strictly on ecology, will nevertheless be of great interest and use to the ecologist, who will frequently need to look at the insect/plant relationship. Each paper has its own bibliography, but there is no general index.

671   Hynes, H.B.N. 1970. *The ecology of running waters.* 555pp. Liverpool University Press; Liverpool.

A large proportion of this book is concerned with insects. Twenty chapters cover the subject, including anatomical and behavioural adaptations. A most useful bibliography is included.

672 Marshall, A.G. 1981. *The ecology of ectoparasitic insects.* 459pp. Academic Press; London and New York.

Described as 'indispensable' in the foreword by Miriam Rothschild, this book attempts to examine the ecology of these insects as a whole, covering both those of importance to man and those of no importance. An excellent introduction to, and a comprehensive treatment of, an interesting and important subject.

673 Matthews, E.G. 1976. *Insect ecology.* 226pp. University of Queensland Press; St Lucia.

Most of the examples used are from the Australian fauna. The work is a useful popular approach to the subject.

674 Mound, L.A. and **Waloff, N.** (Eds). 1978. Diversity of insect faunas. *Symposia of the Royal Entomological Society of London* **9**: 1–204.

This volume presents different approaches to the fundamental and expanding theme of diversity of insect faunas. An examination of the relevant mathematical model, by theoretical biologists; the experimental approach of ecologists; and new data from palaeoecologists are included. Each chapter has its own full bibliography.

675 Southwood, T.R.E. (Ed.). 1968. Insect abundance. *Symposia of the Royal Entomological Society of London* **4**: 1–160.

676 Southwood, T.R.E. 1978. *Ecological methods: with particular reference to the study of insect populations.* 2nd edn. xvi + 524pp. Chapman & Hall; London.

Designed to be of use to those who teach animal ecology. A good text, but its great strength is the references. These are very full and have been updated with the new edition.

677 Young, A.M. 1982. *Population biology of tropical insects.* 511pp. Plenum; New York.

Intended as a comprehensive introduction for advanced undergraduates and for first-year graduate students.

EMBRYOLOGY AND DEVELOPMENTAL STAGES

678 Bounhiol, J.-J. 1980. *Larves et métamorphoses.* 229pp. Presses Universitaires de France; Paris.

Some 70 per cent of this work is given over to the Insecta. Each major division of the book has a short bibliography. There are many illustrations, but they are poorly reproduced.

679 Counce, S.J. and **Waddington, C.H.** 1972. *Developmental systems: insects.* 2 volumes. Academic Press; London.

A number of good review articles, with long bibliographies; however, their use is marred by the fact that none of the references gives any title. The preface admits to the need for further reading, but the user is somewhat hampered by the poor style of referencing, making it difficult for the user to use the bibliographies selectively.

680   Highnam, K.C. 1964. Insect reproduction. *Symposia of the Royal Entomological Society of London* **2**: 1–120.

A selection of nine papers aiming to show the precise state of knowledge of the subject at the time of publication. The papers cover genetics, early stages of insect oogenesis, endocrine relationships in insect reproduction, feeding and nutrition, and the environmental control of sexual maturation in insects. Each chapter has a good bibliography. Although many papers have been published since that time on the subject, this collection of papers gives a good overall summary of the subject.

681   Hinton, H.E. 1981. *Biology of insect eggs.* 3 volumes. Pergamon; Oxford.

The definitive work on insect eggs, with very good illustrations. Volume 1 includes an appreciation of Hinton by T.R.E. Southwood. Volume 3 is the bibliography, and the author and subject indexes. Each family is dealt with in some detail. A complete bibliography of Professor Hinton's published papers 1930–77 rounds off the work.

682   Johannsen, O.A. and **Butt, F.H.** 1941. *Embryology of insects and myriapods: the developmental history of insects, centipedes and millipedes from egg deposition to hatching.* xi + 462pp. McGraw-Hill; New York.

The authors claim the book to be a natural development from lecture courses for undergraduates, and it is a good basic text on insect embryology. It assumes some prior knowledge of elementary taxonomy, anatomy and some general biology. The bibliography is restricted to books and papers dealing with embryology proper, and is of course now very outdated.

683   Lawrence, P.A. (Ed.). 1976. Insect development. *Symposia of the Royal Entomological Society of London* **8**: 1–230.

The papers are divided into two main sections: development of the insect egg and imaginal discs of *Drosophila*. The papers fulfil the usual high standards of these Symposia. All the papers have full and up-to-date bibliographies.

684   Merritt, R.W. and **Cummins, R.W.** (Eds). 1978. *An introduction to the aquatic insects of North America.* viii + 441pp. Kendal/Hunt; Dubuque, Iowa.

The volume deals with aquatic insects in all their stages, but for many groups it is the larval stage that is treated in some detail. Keys are given and the work is fully illustrated. A full bibliography is included.

685   Peterson, A. 1948. *Larvae of insects: an introduction to Nearctic species*; Pt I, *Lepidoptera and Hymenoptera*; Pt II, *Coleoptera, Diptera, Neuroptera, Siphonaptera, Mecoptera, Trichoptera.* E. Brothers; Columbus, Ohio.

Intended as a guide to college students for identifying larvae. Although dealing with Nearctic species, it can be used as groundwork for immature stages generally. The illustrations are very good and well labelled.

EVOLUTION AND PHYLOGENY

686 **Boudreaux, H.B.** 1979. *Arthropod phylogeny with special reference to insects.* viii + 320pp. Wiley Interscience; New York.

A useful exposition of the subject. Of special value as a text for student courses on insect evolution and phylogeny. The volume is well indexed; the bibliography includes a number of references unseen by the author.

687 **Callahan, P.S.** 1972. *The evolution of insects.* 192pp. Holiday House; New York.

A semi-popular work; a readable introduction to the subject for the amateur or beginner.

688 **Cederholm, L.** (Ed.). 1981. Advances in insect systematics and phylogeny. *Entomologica Scandinavica Suppl.* **15**: 1–415.

A comprehensive collection of papers on the subject, by foremost specialists in certain subject areas. Each paper has a useful and up-to-date bibliography.

689 **Gupta, A.P.** (Ed.). 1979. *Arthropod phylogeny.* 762pp. Van Nostrand Reinhold; New York.

A good selection of chapters, each by known authors. A useful compendium of information on this controversial subject. The work has a taxonomic and subject index.

690 **Hennig, W.** 1969. *Die Stammesgeschichte der Insekten.* 436pp. Senckenbergisches Naturforschendes Gesellschaft; Berlin.
Translation: *Insect phylogeny.* 514pp. Translated and edited by A.C. Pont. 1981. John Wiley; Chichester.

The English translation of this much-quoted work makes it now more readily available to a wider audience.

691 **Kettlewell, H.B.D.** 1973. *The evolution of melanism: the study of a recurring necessity, with special reference to industrial melanism in the Lepidoptera.* xxiv + 423pp. Clarendon Press; Oxford.

Standard basic text on the subject. It is not too specialized and can be used by student and research worker alike

692 **Kristensen, N.P.** 1981. Phylogeny of insect orders. *Annual Review of Entomology* **26**: 135–57.

A useful critical review of the subject with a full bibliography.

693 **Laporte, L.F.** 1978. *Evolution and the fossil record.* viii + 222pp. (*Readings from 'Scientific American'.* Collected papers 1949–78.)

A good semi-popular account of the subject. Some of the articles are on a very general level, whilst others are on specific subjects, by authors well known in their subject area.

Pont, A.C. *Insect Phylogeny* – see entry 690

694   Rodendorf, B.B. 1962. *Fundamentals of palaeontology: handbook for palaeontologists of the USSR. Arthropoda.* [In Russian]. 560pp. Izdatel'stvo Akademii Nauk SSSR; Moscow.

A good Russian text for students and researcher alike. The work covers all groups of arthropods, but much of the work is devoted to insects. The illustrations are rather poor.

695   Rodendorf, B.B. and **Rasnitzyn, A.P.** (Eds). 1980. Historical development of the class Insecta. [In Russian]. *Trudy Paleontologicheskogho Instituta* **175**: 1–269.

A controversial work, but difficult to follow unless the user is fluent in Russian. A cover-to-cover translation is in preparation.

See also the section 'Fossil insects' – a number of works on the phylogeny and historical development of particular families of insects are listed.

FLIGHT

696   Candy, D.J. and **Kilby, B.A.** 1978. *Insect biochemistry and function.* 314pp. Chapman & Hall; London.

The first two chapters of the work give a good survey of the biochemistry of fuels required for flight.

697   Dalton, S. 1975. *Borne on the wind: the extraordinary world of insects in flight.* 160pp. Chatto & Windus; London.

Published mainly for its photographs, although the work does have some text. These excellent pictures show more adequately than words the exact movements of wings in flight.

698   Johnson, C.G. 1969. *Migration and dispersal of insects by flight.* xii + 763pp. Methuen; London.

The standard text on the subject. There is a full bibliography at the end of the volume, well cross-referenced in the text.

699   Nachtigall, W. 1968. *Gläserne Schwingen: aus einer Werkstatt biophysikalischer Forschung.* 158pp. Heinz Moos Verlag; Munich.

Written for both layman and student. In a complex subject area, the work is clearly written and defined, and well illustrated. Entry 700 is an English translation.

700   Nachtigall, W. 1974. *Insects in flight: a glimpse behind the scenes in biophysical research.* Translated by H. Oldroyd, R. H. Abbott and M. Blederman-Thorson. 153pp. George Allen & Unwin; London.

A translation of entry 699.

701   Pringle, J.W.S. 1957. *Insect flight.* viii + 133pp. Cambridge University Press; Cambridge, England.

A good student text and introduction to the subject. Much has since been published on the subject, but this is still a useful entry to the subject.

702 Rainey, R.C. (Ed.). 1976. Insect flight. *Symposia of the Royal Entomological Society of London* **7**: 1–287.

Collection of papers on flight read at a symposium in 1974. Each paper has its own up-to-date bibliography.

703 Tregear, R.T. (Ed.). 1979. *Insect flight muscle.* 367pp. North-Holland; Amsterdam.

A collection of papers read at an Oxford symposium in 1977. A very specialized subject area, but a most useful collection of papers with good bibliographies.

FOSSIL INSECTS

704 Bode, A. 1952. Die Insektenfauna des Ostniedersächsischen oberau Lias. *Palaeontographica A* **103**: 1–375.

Covers most of the groups of insects. Many illustrations, but poorly reproduced.

705 Bolton, H. 1921–2. *A monograph of the fossil insects of the British coal measures.* 156pp. Palaeontographical Society; London.

Now out of date, but it is the only work of its kind.

706 Callahan, P.S. 1972. *The evolution of insects.* 192pp. Holiday House; New York.

Written in a fairly popular style, and a nice introduction for the student or amateur.

707 Carpenter, F.M. 1971. Adaptations among Paleozoic insects. *Proceedings of the North American Paleontological Convention* **1**: 1236–51.

A short paper, but of fundamental importance in the subject.

708 Coope, G.R. 1979. Late Cenozoic fossil Coleoptera: evolution, biogeography and ecology. *Annual Review of Ecology and Systematics* **10**: 247–67.

A short, pertinent review of the subject with a good up-to-date bibliography.

709 Crowson, R.A. *et al.* 1967. Arthropoda: Chelicerata, Pycnogonida, Palaeoisopus, Myriapoda and Insecta. In *Fossil record* Chapter 19: 499–534. Geological Society; London.

Gupta, A.P. *Arthropod phylogeny – see* entry 689

710 Handlirsch, A. 1906–8. *Die fossilen Insekten und die Phylogenie der rezenten Formen. Ein Handbuch für Paläontologen und Zoologen.* ix + 1430pp. W. Engelmann; Leipzig.

See entry 711.

711 Handlirsch, A. 1937–9. Neue Untersuchungen über die fossilen Insekten mit Ergänzungen und Nachträgen sowie Ausblicken auf phylogenetische, palaeographische und allgemein biologische Probleme. *Annalen Naturhistorischen Museums Wien* **48**: 1–140; **49**: 1–240.

The 1906 text (entry 710) and its little-known supplement of 1937–9 are now dated. However, Handlirsch was the first to bring together and publish a great compilation on fossil insects, and his book can still be used as a general text on the subject.

712   Handlirsch, A. 1919. Revision der paläozoischen Insekten. *Denkschriften der Akademie der Wissenschaften in Wien, Mathematisch-Naturwissenschaftliche Klasse* **96**: 1–82.

713   Heie, O.E. 1967. Studies on fossil Aphides (Homoptera: Aphidoidea), especially in the Copenhagen collection of fossils in Baltic amber. *Spolia Zoologica. Musei Hauniensis* **26**: 1–273.

Although the paper is basically written on the collections in Copenhagen, there is much additional information on fossil aphids in general.

Hennig, W. *Die Stammesgeschichte der Insekten* (*Insect phylogeny*) – *see* entry 690

714   Kusnezov, N.J. 1941. *A revision of the amber Lepidoptera*. [In Russian]. 136pp. Izdatel'stvo Akademii Nauk SSSR; Moscow.

The standard work on the subject, with a good bibliography.

715   Laporte, L.F. 1978. *Evolution and the fossil record.* viii + 222pp. (*Readings from 'Scientific American'*. Collected papers 1949–78.)

Some general articles on evolution, others more specifically on fossil insects, originally written for the journal *Scientific American*. They give a good semi-popular introduction to the subject. As usual with the articles in this periodical, they are very well illustrated.

716   Larsson, S.G. 1978. Baltic amber – a palaeobiological study. *Entomonograph* **1**: 1–192.

Some 90 per cent of the work covers the insects. Most of the insect families are covered and each chapter has its own bibliography.

717   Laurentiaux, D. 1953. In Piveteau, J. (Ed.). *Traité de paléontologie*; Vol. 3, *Arthropodes*, pp. 397–527. Masson; Paris.

These pages cover the Insecta. The beginning of the chapter deals with the history of the subject. Individual orders are dealt with under separate headings.

Pont, A.C. *Insect phylogeny* – *see* entry 690

718   Popov, Y.A. 1971. Historical development of the Hemiptera. [In Russian]. *Trudy Paleontologicheskogho Instituta* **129**: 1–228.

Although a standard work for the Hemiptera, it has no English or German summary, and a good reading knowledge of Russian is needed for its study.

719   Rasnitsyn, A.P. 1969. Origin and evolution of lower Hymenoptera. [In Russian]. *Trudy Paleontologicheskogho Instituta* **123**: 1–183. Translated in 1979 for the USDA Agricultural Research Service by America Publications, New Delhi.

A most useful work for the evolution of the lower Hymenoptera. Difficult to use for non-Russian scholars, but now more readily available for study, thanks to the cover-to-cover translation.

720   Riek, E.F. 1970. Fossil history. In *Insects of Australia.* Chapter 8: pp. 168–86. CSIRO; Canberra.

Although the examples are from the Australian fauna, this is a most useful, if brief, general account of fossil history; many students and general readers will find it a helpful introduction.

721   Rodendorf, B.B. *et al.* 1964. Palaeozoic insects of the Kuzbacs. [In Russian]. *Trudy Paleontologicheskogho Instituta* **85**: 1–705.

Another complete review in this Russian series. New taxa are described and keys for identification are provided. The work is well illustrated and indexed. There is no English or German summary.

722   Rodendorf, B.B. 1962. *Fundamentals of palaeontology. A handbook for palaeontologists of the USSR. Arthropoda.* [In Russian]. 560pp. Izdatel'stvo Akademii Nauk SSSR; Moscow.

A most comprehensive textbook.

723   Rodendorf, B.B. 1964. Historical development of the Diptera. [In Russian]. *Trudy Paleontologicheskogho Instituta* **100**: 1–311.

A cover-to-cover translation of this work is available: Moore, J.E. and Thiele, I. (Translators) (1974). *Historical development of the Diptera.* xv + 360pp. University of Alberta Press; Edmonton.

724   Rodendorf, B.B. (Ed.). 1980. Historical development of the Insecta. *Trudy Paleontologicheskogho Instituta* **175**: 1–269.

This work is a collection of papers by several authors of considerable experience and authority on fossil insects. It is a somewhat controversial work, since it departs from current practice in the nomenclature of taxa in the order–class group. The work is entirely in Russian, but a very full and explanatory review in English can be found in the scientific periodical *Entomologia Generalis* **7**: 105–108. It is hoped that a cover-to-cover translation will appear in the not too distant future.

725   Scudder, S.H. 1875. *Fossil butterflies.* 99pp. Memoirs of the American Association for the Advancement of Science; Washington, DC.

Although an old work, it is still frequently referred to.

726   Sharov, A.G. 1968. Phylogeny of the Orthopteroidea. [In Russian]. *Trudy Paleontologicheskogho Instituta* **118**: 1–218.

A cover-to-cover 1971 translation by the Israel Program for Scientific Translation is available.

727   Wooton, R.J. 1981. Palaeozoic insects. *Annual Review of Entomology* **26**: 319–44.

A most useful 'state of the art' review. As with most papers published in this review series, there is a good, complete and up-to-date bibliography.

728   Zeuner, F. 1939. *Fossil Orthoptera Ensifera.* 321pp. British Museum (Natural History); London.

GENETICS (*see also* Cytogenetics)

729   Ashburner, M. and **Novitski, E.** (Eds). 1976–. *The genetics and biology of* Drosophila. 9 volumes (to date). Academic Press; New York.

Studies with *Drosophila* have contributed much to the understanding of heredity and are frequently used in experimental genetics research. This work is therefore of value to advanced students of genetics as well as those involved in *Drosophila* research. The work should be complete in 12 volumes, the last of which is planned to include a consolidated index.

730   Ewing, A.W. and **Manning, A.** 1967. The evolution and genetics of insect behaviour. *Annual Review of Entomology* **12**: 471–94.

A good broad review article, with a full bibliography up to 1966.

731   Ford, E.B. 1975. *Ecological genetics.* 4th edn. xx + 442pp. Chapman & Hall; London.

The classic account of genes in populations, with insects featuring prominently.

732   Herskowitz, I.H. 1977. *Principles of genetics.* 2nd edn. 836pp. Macmillan; New York.

'A definitive synthesis of all aspects of genetic knowledge, with special emphasis on molecular genetics.' A teaching text for undergraduates. The literature cited after each chapter includes general references to the work in the chapter, followed by references to specific numbered sections. On pp. 717–34 is a glossary of terms used in the text. The book has a good subject index.

733   King, R.C. (Ed.). 1975. *Handbook of genetics*; Vol. 3, *Invertebrates of genetic interest.* 874pp. Plenum; London.

The volume is almost entirely on insects. Each chapter has a bibliography of cited literature. An author and subject index is included.

734   Lewontin, R.C. 1974. *The genetic basis of evolutionary change.* 346pp. Columbia University Press; New York.

An important treatise on population genetics and evolution.

735   Remington, C.L. 1968. The population genetics of insect introduction. *Annual Review of Entomology* **13**: 415–26.

A brief but useful review of the subject, with a good bibliography.

736 Robinson, R. 1971. *Lepidoptera genetics.* ix + 687pp. Pergamon; Oxford.

A review of the work done on genetics of the Lepidoptera.

737 Wagner, R.P. and **Selander, R.K.** 1974. Isozymes in insects and their significance. *Annual Review of Entomology* **19**: 117–38.

A useful summary of the applications of electrophoresis in entomology.

738 Wright, J.W. and **Pal, R.** 1967. *Genetics of insect vectors of disease.* xx + 794pp. Elsevier; Amsterdam.

A useful bibliography is included, but is now somewhat outdated.

MORPHOLOGY

739 Chapman, R.F. 1982. *The insects: structure and function.* 3rd edn. 919pp. Hodder & Stoughton; London.

A widely used, reasonably priced and very comprehensive student textbook. As in previous editions, the majority of the figures are reproduced from standard works but the quality of reproduction is much improved. A useful bibliography completes each chapter.

740 Manton, S.M. 1977. *The Arthropoda: habits, functional morphology and evolution.* xx + 527pp. Clarendon Press; Oxford.

A good reference text, bringing together much of the recent work on this subject.

741 Manton, S.M. 1949–72. The evolution of Arthropodan locomotary mechanism. [In 10 parts.] *Journal of the Linnean Society (Zoology)* **41**: 529–71; **42**: 93–117; **42**: 118–67; **42**: 299–368; **43**: 153–87; **43**: 487–556; **44**: 383–461; **46**: 251–500; **46**:103–41; *Zoological Journal of the Linnean Society* **51**: 203–400.

This important series of papers is a complete review of the subject, covering all the arthropod groups including the insects.

742 Matsuda, R. 1965. Morphology and evolution of the insect head. *Memoirs of the American Entomological Institute* No. **4**: 1–334.

See entry 744.

743 Matsuda, R. 1970. Morphology and evolution of the insect thorax. *Memoirs of the Entomological Society of Canada* No. **76**: 1–431.

See entry 744.

744 Matsuda, R. 1976. *Morphology and evolution of the insect abdomen, with special reference to developmental patterns and their bearings on systematics.* viii + 534pp. Pergamon; Oxford.

This and the previous two works make a good collection of working texts on insect morphology of use both to the student and the research worker. All have good bibliographies.

745   Niceville, A.C. (Ed.). 1970. Insect ultrastructure. *Symposia of the Royal Entomological Society of London* **5**: 1–185.

The papers fall largely into two main categories: the exoskeletal system and the control mechanisms. Special attention is given to interrelating structure and function. The work is well illustrated and has useful bibliographies.

746   Snodgrass, R.E. 1935 *Principles of insect morphology.* 667pp. McGraw-Hill; New York.

Although now somewhat dated, this book is still regarded as *the* standard text. It is a good introduction to the subject. One of its major assets is the glossary appended to each chapter. There is a large bibliography at the end of the volume, but this is of course very dated.

747   Usherwood, P.N.R. (Ed.). 1975. *Insect muscle.* 621pp. Academic Press; London.

A full and comprehensive account of the work done in recent years involving work with insect muscle. Twelve contributors' papers are brought together in this well illustrated volume.

PATHOLOGY

748   Aoki, K. 1957. *Insect pathology.* [In Japanese]. 493pp. Gihûdô Press; Tokyo.

Well regarded as a text on pathology, for those who read Japanese. The work has a good index and a very full bibliography.

749   Cantwell, G.E. (Ed.). 1974. *Insect diseases.* 2 volumes. Marcel Dekker; New York.

The work is aimed at the postgraduate student. No claim is made for a complete bibliography, but some 900 entries are included.

750   Gibbs, A.J. 1973. *Viruses and invertebrates.* 673pp. North-Holland; Amsterdam. (North-Holland Research Monographs, *Frontiers of Biology*, Vol. 31.)

The three main parts cover the insect groups as plant virus vectors; ecology of viruses; and viruses and particular invertebrates and their control.

751   Krieg, A. 1961. *Grundlagen der Insektenpathologie Viren-, Rickettsien- und Bakterien-Infektionen.* 304pp. Dietrich Steinkopff; Darmstadt.

A good early text, but the subject has moved ahead somewhat since this work was published.

752   Maramorosch, K. and **Shope, R.E.** 1975. *Invertebrate immunity. Mechanisms of invertebrate vector–parasite relations.* 365pp. Academic Press; London.

'The book provides the first modern, integrated presentation of phenomena and mechanisms pertaining to immunity in invertebrate animals.' Of most use to research workers and graduate students.

753 Poinar, G.O. and **Thomas, G.M.** 1978. *Diagnostic manual for the identification of insect pathogens.* 218pp. Plenum Press; London.

Valuable for diagnosing the microbial diseases of insects. Pathogens included are fungi, protozoans, bacteria, viruses and rickettsias. Primarily an identification guide; general background information on the pathogens is also included.

754 Smith, K.M. 1976. *Virus–insect relationships.* 291pp. Longman; London.

The volume is divided into two main sections: types of insect virus disease and aspects of the study of virus–insect relationships. The work has a good bibliography and is well indexed.

755 Steinhaus, E.A. (Ed.). 1963. *Insect pathology: an advanced treatise.* 2 volumes. Academic Press; London.

Much wider in scope than the first edition of this work by the same editor. Seventeen invited authors cover the subject in some detail. An authoritative work for the research scholar.

756 Weiser, J. 1977. *An atlas of insect diseases.* 81pp. 240 plates. Dr W. Junk; The Hague.

A series of photographs for the entomologist and insect pathologist, representing all types of insect pathogens.

PHYSIOLOGY

757 Beck, S.D. 1980. *Insect photoperiodism.* 2nd edn. 387pp. Academic Press; London and New York.

An update of the earlier edition of 1968. A full bibliography, but no titles to the cited papers, thus giving the reader no clue as to their content.

758 Brues, C.T. 1946. *Insect dietary: an account of the food habits of insects.* 466pp. Harvard University Press; Cambridge, Massachusetts.

The work covers most aspects of feeding habits such as galls, blood-sucking insects, herbivores and predatory insects.

759 Candy, D.J. and **Kilby, B.A.** 1975. *Insect biochemistry and function.* 314pp. Chapman & Hall; London.

A useful volume for both graduates and undergraduates. The first two chapters give a good survey of the biochemistry of fuels required for flight. The work was also issued in a paperback edition in 1978.

760 Dethier, V.G. 1963. *The physiology of insect senses.* 266pp. Methuen; London.

Mainly a student text, the work covers the subject area in some detail.

761 Gilmour, D. 1965. *The metabolism of insects.* 195pp. Oliver & Boyd; Edinburgh.

A somewhat brief account of this complex subject area. The subject is usually well covered in larger, more comprehensive works on physiology, or in the multi-volume works. However, this is a useful student text in one volume.

762   Heinrich, B. (Ed.). 1981. *Insect thermoregulation.* 328pp. John Wiley; Chichester.

An up-to-date summary and review of the work that has been done on thermoregulation. The papers were originally given at a Symposium of the American Society of Zoologists. The book has very good bibliographies and is well indexed.

763   Miller, T.A. (Ed.). 1979. *Insect neurophysiological techniques.* 308pp. Springer-Verlag; Berlin. (Springer Series in Experimental Entomology.)

The work is a detailed study of the techniques used in this complex research area. Part 1 describes the instruments and materials useful in this kind of research, part 2 the methods used to study unrestrained insects and part 3 tethered insects. Organ and tissue preparation is described in part 4. A list of equipment suppliers is appended.

764   Miller, T.A. (Ed.). 1980. *Neurohormonal techniques in insects.* 282pp. Springer-Verlag; Berlin. (Springer Series in Experimental Entomology.)

The work discusses the current neurochemical research on neurohormones. Although insects are used as subjects the work has broader applications. Of most use to research scholars.

765   Miller, T.A. (Ed.) 1980. *Cuticle techniques in arthropods.* 410pp. Springer-Verlag, Berlin. (Springer Series in Experimental Entomology.)

This work will be of use to both the entomologist and the comparative physiologist. It is a detailed study of the techniques used to study the cuticle of insects and other arthropods. Each chapter has its own bibliography. Addresses of equipment suppliers are also included.

766   Novak, V.J.A. 1975. *Insect hormones.* 2nd edn. 600pp. Chapman & Hall; London.

Although innumerable papers have been published on specific aspects of research on this subject, there is much basic and useful information in this work. This classic text, originally published in German, has gone through four editions. The bibliography is very full and has been updated with each new edition.

767   Rockstein, M. (Ed.). 1978. *Biochemistry of insects.* 649pp. Academic Press; London.

The book is designed to serve as a basic textbook in the field, and most of the material is written as a teaching aid. Some chapters are at a more advanced level. The references are very full and some chapters offer them at two levels: one for the undergraduate and the other for the more advanced student or research scientist.

768 **Rockstein, M.** (Ed.) 1973–4. *The physiology of the Insecta.* 2nd edn. 6 volumes. Academic Press; London.

A completely revised edition with many additional chapters. The bibliographies are full but poorly set out. The references give only the first page number and no paper title, so that the user has no idea of the size or possible scope of the paper referred to.

769 **Strausfeld, N.J.** and **Miller, T.A.** 1980. *Insect neuroanatomical techniques: insect nervous system.* 496pp. Springer-Verlag; Berlin. (Springer Series in Experimental Entomology.)

Techniques are described for the preparation of tissue for whole mounts and for light and electron microscopy. There is a great deal of new information on techniques and recent developments in methods of identifying various types of nerves. There is one set of references for the whole volume.

770 **Wigglesworth, V.B.** 1972. *Principles of insect physiology.* 7th edn. 827pp. Chapman & Hall; London.

A standard text for many years and through all its editions. This newest edition is updated in parts and the bibliography brought up to date. The volume is easy to use and is well illustrated. A very good index is included.

VISION

771 **Bernhard, C.G.** (Ed.). 1966. *The functional organization of the compound eye.* 591pp. Pergamon Press; London. (Proceedings of an International Symposium, Stockholm, 1965.)

Many of the papers included are on the work done with the compound eye of insects. The main sections include optics, rhabdon structure, photochemistry, receptor excitation and integration of visual input. Of most use to the research scientist.

772 **Frantsevich, L.I.** 1980. *Optical analysis of space in insects.* [In Russian.] Naukova Dumka; Kiev.

All groups of insects are covered in this work. The anatomy and physiology of insect optics are discussed and illustrated, although the illustrations are rather poor. There is a full bibliography of 1083 references clearly set out and well up to date.

773 **Horridge, G.A.** (Ed.). 1975. *The compound eye and vision of insects.* 595pp. Clarendon Press; Oxford.

A collection of papers from a symposium at the International Congress of Entomology, Canberra. The papers are contributions to research rather than 'state-of-the-art' reviews.

774 **Mazokhin-Porshnyakov, G.A.** 1969. *Insect vision.* 306pp. Plenum Press; New York.

Translated from the Russian by R. and L. Masironi, with some updating.

Economic entomology

MEDICAL AND VETERINARY ENTOMOLOGY

775 Abul-Hab, J.K. 1979. *Medical and veterinary entomology in Iraq.* [In Arabic]. 2 volumes. University of Baghdad; Baghdad.

Latin specific names are included in the volumes. Several photographs are included in vol. 2, but they are poorly reproduced.

776 Bates, M. 1949. *The natural history of mosquitoes.* 379pp. Macmillan; New York.

Lightly but authoritatively written; a good introduction to the subject.

777 Bettini, S. (Ed.). 1978. *Arthropod venoms.* 977pp. Springer-Verlag; Berlin.

Provides a very thorough coverage of the subject, including pathology, epidemiology, treatment, etc. The volume is well indexed and each chapter has a good bibliography.

778 Blum, M.S. 1981. *Chemical defenses of arthropods.* 562pp. Academic Press; New York.

The first detailed treatment of the chemical ecology of arthropod defences, covering scorpions, spiders and other arthropods, including various insects. Catalogues all the compounds evolved by arthropods for defensive utilization. A useful but highly technical work.

779 British Museum (Natural History). Economic Series: No. 2a (1969), *Lice* (24pp). No. 5 (1973), *The bed-bug* (17pp).

Two of a useful series of pamphlets published by the British Museum (Natural History). They are well illustrated. Suggestions for control are offered. The pamphlets are updated at frequent intervals.

780 Busvine, J.R. 1976. *Insects, hygiene and history.* 262pp. Athlone; London.

A very readable account of the insect pests of man's social habits and hygiene (or lack of it!). Written in a semi-popular style, this volume contains a great deal of useful background information. It is well illustrated, and includes 322 references and an index.

781 Busvine, J.R. 1980. *Insects and hygiene: the biology and control of insect pests of medical and domestic importance.* 3rd edn. 568pp. Chapman & Hall; London.

A classic work, well illustrated with good, clear line drawings.

782 Buxton, P.A. 1955. *The natural history of tsetse flies: an account of the biology of the genus* Glossina *(Diptera).* 816pp. H.K. Lewis; London. (London School of Hygiene and Tropical Medicine Memoir No. 10.)

Still the standard teaching text for *Glossina.* The subject is covered in some detail; it is set out in a clear and easy-to-use style. The work is well illustrated and indexed.

783 Centre for Overseas Pest Research. 1978. *Note on allergy to locusts.* 9pp. Centre for Overseas Pest Research; London.

This booklet outlines the origin of the allergic reactions that may occur among those working with locusts, and indicates the measures that may be taken to prevent them.

784 Cloudsley-Thompson, J.L. 1976. *Insects and history.* 242pp. Weidenfeld & Nicolson; London.

Written in a semi-popular style, this work is a useful survey of the historical importance of insects, particularly as carriers of disease. A brief but up-to-date bibliography is included.

785 Davies, H. 1977. *Tsetse flies in Nigeria: a handbook for junior control staff.* 3rd edn. 340pp. Oxford University Press; Ibadan.

Aimed, as its title suggests, primarily at tsetse workers rather than entomologists, this simply written book provides the non-specialist reader with an excellent, concise introduction to the subject.

786 Derbeneva-Ukhova, V.P. (Ed.). 1974. *Guide to medical entomology.* [In Russian]. 360pp. 'Meditsina'; Moscow.

A well illustrated work covering groups of insects of medical importance. Gives details of their biology and distribution. The references are divided by subject at the end of the volume.

787 Eichler, W. 1980. *Grundzüge der veterinärmedizinischen Entomologie: ausgewählte Beispiele wichtiger Parasitengruppe.* 184pp. VEB Gustav Fischer Verlag; Jena.

Reviews the taxonomy and general characters, development and biology, and the recognition and determination of insects of veterinary and medical importance. Well illustrated both by line drawings and by photographs.

788 Forattini, O.P. 1962–5. *Entomologia médica.* 4 volumes. Faculdade de Higiene e Saúde Pública, Universidade de São Paulo; São Paulo.

A strong emphasis on the Neotropical region. The work includes full accounts of the insects of medical importance with keys for their identification, plus a full bibliography.

789 Frazier, C.A. 1969. *Insect allergy: allergic and toxic reactions to insects and other arthropods.* 493pp. Warren H. Green; St Louis, Missouri.

Many of the insects of the world cause allergic reactions in man, both by stinging and urticaria. These are all dealt with under the headings of the various groups of insects. The work is well illustrated, and fully indexed. The referencing is poor.

790 Frazier, M.D. and **Brown, F.K.** 1980. *Insects and allergy and what to do about them.* 272pp. University of Oklahoma Press; Norman, Oklahoma.

The text focuses on arthropods that commonly pose health problems in the US and the prevention or treatment of allergic and toxic reactions to bites, stings and other types of exposure. The work is poorly referenced.

791 Furman, D.P. and **Catts, E.P.** 1982. *Manual of medical entomology.* 4th edn. 207pp. Cambridge University Press; London.

Intended as a teaching tool to familiarize students with the use of taxonomic keys for the identification of arthropods of medical or veterinary importance. Also includes a limited amount of information on collection and preparation, and on laboratory rearing.

792 Gillett, J.D. 1971. *Mosquitos [sic].* 274pp. Weidenfeld & Nicolson; London.

A good general introduction to the mosquitoes. Written in an easy-to-follow style, the work is well illustrated and has a full bibliography divided by subject. A short glossary is given on pp. 241–3.

793 Greenberg, B. 1971. *Flies and disease.* 2 volumes. Princeton University Press; Princeton, New Jersey.

Volume 1 deals with ecology, classification and biotic associations; vol. 2 with biology and disease transmission. This is a very full and comprehensive work. Each volume has an extensive bibliography and vol. 1 includes useful identification keys and some fine coloured plates.

794 Harwood, R.F. and **James, M.T.** 1979. *Entomology in human and animal health.* 7th edn. 548pp. Macmillan; New York.

Authoritative and comprehensive work covering all aspects of medical and veterinary entomology. Formerly published as *Herm's Medical entomology*, this is a well-established standard student text. Well illustrated and contains an extensive bibliography.

795 Horsfall, W.R. 1972. *Mosquitoes: their bionomics and relation to disease.* 723pp. Hafner; New York.

The object of the book is to summarize, as completely as possible, the large and varied literature on mosquitoes that pertains to their bionomics and relation to disease. The work is set out systematically according to genus and species. On pp. 603–7 is a brief glossary of terms useful to the various specialists (doctors, entomologists, ecologists, etc.) who may be involved in the subject. Some 80 pages of references are included, but references give only journal titles, without the titles of the papers. The volume is well indexed.

796 Hyneman, D., Hoogstral, H. and **Djigounian, A.** 1980. *Bibliography of leishmania and leishmanial diseases.* 2 volumes. US Naval Medical Research Unit. No. 3; Cairo.

Lists the literature of the epidemiology, biology, systematics and clinical aspects of leishmanial agents of human disease and their phlebotomine vectors.

797 Laird, M. (Ed.). 1977. *Tsetse: the future for biological methods in integrated control.* 220pp. International Development Research Centre; Ottawa.

Far more comprehensive than its title suggests, this work provides a great deal of basic information on tsetse biology, ecology, field techniques, laboratory rearing and control. Illustrated with photographs, and contains a lengthy bibliography.

798 Laird, M. (Ed.) 1981. *Blackflies: the future for biological methods in integrated control.* xii + 399pp. Academic Press; London.

Similar in style and content to the work described above.

799 Leclercq, M. 1969. *Entomological parasitology: the relations between entomology and the medical sciences.* 158pp. Pergamon; Oxford.

A useful and concise treatment of this important subject. Contains sections on edible insects, and on forensic entomology. Each section concludes with a short bibliography.

800 Levine, M.I. and **Lockey, R.F.** 1981. *Monograph on insect allergy.* 84pp. Typecraft; Pittsburgh, Pennsylvania.

A brief, up-to-date, authoritative source of information on insect allergies. Diagnosis and treatments are dealt with and there is a chapter on protective measures against insect stings.

801 Mattingly, P.F. 1969. *The biology of mosquito-borne disease.* xii + 184pp. George Allen & Unwin; London.

A most succinct account of this important subject. The references are at the end of each chapter and include only broad review articles.

802 Morishita, T. and **Kano, R.** (Eds). 1967. *Medical parasitology.* [In Japanese]. 412pp. Nanzando; Tokyo.

Includes most groups of insects of medical importance. Latin names are used throughout. The volume is well indexed, using Latin and Japanese names. It is poorly illustrated.

803 Mulligan, H.W. (Ed.). 1970. *The African trypanosomiases.* 950pp. George Allen & Unwin; London.

Sponsored by the UK's Ministry of Overseas Development, this is the standard work in its field. Numerous authors have contributed sections that collectively cover almost every aspect of *Glossina* and the African trypanosomiases. There is an 82-page bibliography that includes virtually all the more significant earlier *Glossina* papers.

804 Nash, T.A.M. 1969. *Africa's bane: the tsetse fly.* 224pp. Collins; London.

Written for the general reader rather than the entomologist, this book provides a good general introduction to *Glossina*, covering historical background, distribution, ecology, biology and control. There are numerous black-and-white photographs and a short, selective bibliography.

805 Neveu-Lemaire, M. 1938. *Traité d'entomologie médicale et vétérinaire.* Vigot Frères; Paris.

All the insect groups are covered in some detail. The volume is well indexed, but has few references, most of which are now out of date. The main body of the work is a good standard text.

806 Sasa, M. 1976. *Human filariasis. A global survey of epidemiology and control.* 819pp. University Park Press; Tokyo.

A most useful text on the subject; the work covers insects as vectors of disease. A full bibliography is included; unfortunately there is no index to the volume.

807 Sasa, M. *et al.* (Ed.). 1977. *Animals of medical importance in the Nansei Islands in Japan.* Shinjuku Shobo; Tokyo.

The area covered includes two important zoogeographical zones. All the major insects of medical importance are included; brief accounts of life histories are given and some keys for identification.

808 Service, M.W. 1976. *Mosquito ecology: field sampling methods.* 583pp. Applied Science Publishers; London.

This is the first comprehensive work on the many different methods available for collecting culicids, and the problems associated with sampling populations. Although basically concerned with mosquitoes, many of the traps and procedures described are applicable to the study of other medically important insects. An enormous number of references are listed, for which a separate author index is included.

809 Service, M.W. 1980. *A guide to medical entomology.* 226pp. Macmillan; London.

Written for those working in the fields of medicine, hygiene, public health or parasitology who have no specialized knowledge of medical entomology. It provides basic information on the recognition of arthropod vectors of disease, their biology and life cycles, and their role in the transmission of diseases to man. It also presents a guide to their control. Good, clear line drawings throughout, and a fairly extensive bibliography.

810 Smith, K.G.V. (Ed.). 1973. *Insects and other arthropods of medical importance.* 561pp. British Museum (Natural History); London.

An excellent, comprehensive work intended primarily for the identification of insects and other arthropods of importance in human medicine. Each of the 16 contributors is a specialist on the taxonomy of the group concerned. Contains considerable information on the role of disease vectors and their biology, and all groups are treated on a world basis. There are extensive and useful bibliographies, numerous line drawings and 12 plates.

811 Soulsby, E.J.L. 1971. *Helminths, arthropods and protozoa of domesticated animals.* 6th edn. xix + 824pp. Baillière, Tindall & Cassell; London.

This excellent textbook, the sixth edition of Mönnig's *Veterinary helminthology and entomology*, has a large section on arthropods. Each section has its own bibliography; a host–parasite list and an index are included.

812 West, L.S. and **Peters, O.B.** 1973. *An annotated bibliography of Musca domestica Linnaeus.* 743pp. Dawson; Folkestone.

Covers every aspect of the housefly from the earliest references up to 1969. The main bibliography, comprising some 5720 references, is split into three chronological sections, while a fourth deals with publications on research techniques. Within each section, references are arranged alphabetically, and there is a very useful overall subject index. The only comprehensive work of its kind, extremely well organized, and particularly useful in that each bibliographic reference is accompanied by a brief summary of content if this is not clear from the title.

813 World Health Organization. 1975. *Manual of practical entomology.* 2 parts. WHO; Geneva. (Part 1, *Vector bionomics and organization of antimalaria activities*; part 2, *Methods and techniques.*)

The work is aimed at both the practical entomologist and the malariologist, in a malaria programme. As with most WHO documents the subject is well covered. It has some good illustrations, and a useful bibliography.

814 World Health Organization. 1976. *Epidemiology of onchocerciasis.* 94pp. WHO; Geneva. (Technical Report Series No. 597.)

A succinct account of the subject.

815 Wright, J.W. and **Pal, R.** (Eds). 1967. *Genetics of insect vectors of disease.* 794pp. Elsevier; Amsterdam.

Provides a comprehensive and systematic review of this subject, illustrated with many photographs. Each chapter concludes with a fairly lengthy bibliography, although there is no overall author index.

816 Yasarol, S. 1978. *Medikal parazitologi.* 444pp. Ege Üniversitesi; Izmir.

A fairly comprehensive text, and the only work of its kind published in Turkish. Includes a rather limited bibliography; illustrations poor.

817 Zahar, A.R. 1979–80. *Studies on leishmaniasis vectors/reservoirs and their control in the Old World.* 4 parts. WHO; Geneva.

The series is published in the WHO/VBC series. These are not generally regarded as 'published' works, but they are excellent sources of current information.

818 Zumpt, F. 1973. *The stomoxyine biting flies of the world. Diptera: Muscidae. Taxonomy, biology, economic importance and control measures.* 175pp. Gustav Fischer Verlag; Stuttgart.

An authoritative review of this important subfamily, illustrated with line drawings. Contains a fairly extensive bibliography.

FORENSIC ENTOMOLOGY

819 Easton, A. M. and **Smith, K.G.V.** 1970. The entomology of the cadaver. *Medicine, Science and the Law* **1970**: 208–15.

Includes a useful but short bibliography and five plates of line drawings of species mentioned in the text.

820   Lerclercq, M. 1978. *Entomologie et médecine légale: datation de la mort.* 100pp. Masson; Paris.

A short and somewhat gruesome illustrated book on forensic entomology. Includes a useful bibliography of 111 references.

821   Megnin, P. 1894. *La faune des cadavres: application de l'entomologie à la médecine légale.* 214pp. Encyclopédie Leauten; Paris.

An old but very complete work not superseded except perhaps by some papers published in scientific journals.

822   Nuorteva, P. 1977. Sarcophagous insects as forensic indicators. In Tedeschi, C.G. *et al.* (Eds). *Forensic medicine – a study in trauma and environmental hazards.* Vol. 2, pp. 1072–95. W.B. Saunders; Philadelphia.

A detailed study including a very full bibliography of some 176 references.

INSECTS IN AGRICULTURE, FORESTRY AND HORTICULTURE

823   Anderson, J.F. and **Kaya, H.K.** (Eds). 1976. *Perspectives in forest entomology.* 428pp. Academic Press; New York and London.

Aims 'to bring current knowledge on forest pest management into perspective'. Sections include 'The forest', 'Insect bionomics', 'Insect behavior', 'Biological and integrated control' and an appendix that features a useful review of the history of entomology in relation to the US State Agricultural Experiment Stations.

824   Balachowsky, A.S. 1962–. *Entomologie appliquée à l'agriculture.* 4 volumes (to date: *Lépidoptères* (1 and 2); *Coléoptères* (1 and 2)). Masson; Paris.

A very comprehensive and authoritative work. The volumes are well illustrated and cover the insect families in some detail. A full description of each pest is given, including its biology and distribution. Brief details on the control of the pest in question are given. Many photographs show the type of damage caused by the pest to plants.

825   Bedford, E.C.G. (Ed.). 1978. Citrus pests in the Republic of South Africa. *Science Bulletin, Department of Agriculture and Technical Services, Republic of South Africa* **391**: 1–253.

The work is arranged under particular pests. Each specific pest is described with literature references. Distribution and economic importance, along with life history, are discussed in some detail. Control measures are also covered. On pp. 223–5 is a list of common names under English–Afrikaans, Afrikaans–English. A list of host plants of citrus pests listed under scientific names with equivalent English and Afrikaans common names is given.

826   Beeson, C.F.C. 1961. *The ecology and control of the forest insects of India and the neighbouring countries.* 1007pp. Government of India Press; Delhi.

A reprint of the 1941 edition published by the author. It is somewhat out of date, but still a standard work on this subject.

827 Brain, C.K. 1929. *Insect pests and their control in South Africa.* viii + 468pp. Die Nasionale Pers Beperk; Cape Town.

Although now very out of date, this is probably the only comprehensive work for this area, covering both agricultural and medical/veterinary pests. Includes quite a number of line drawings and some black-and-white photographs.

828 Buczacki, S. and **Harris, K.** *Collins Guide to the pests, diseases and disorders of garden plants.* 512pp. Collins; London.

A practical guide to the pests, diseases and disorders that commonly affect fruit, vegetables and ornamental plants growing in gardens in the UK and northern Europe. Information is given concerning the symptoms, biology and treatment for each species, and colour plates are provided as an aid to identification.

829 Buhr, H. 1964. *Bestimmungstabellen der Gallen (Zoo- und Phytocecidien) an Pflanzen Mittel- und Nordeuropas.* 2 volumes. VEB Gustav Fischer Verlag; Jena.

Full identification keys are given. *The* standard text on the subject. The work includes very few illustrations and is of a very technical nature. A full bibliography is included and a good index concludes the work.

830 Chen Shi-Xiang and **Xie Yun-Zhen** (Eds). 1959–. *Economic insect fauna of China.* [In Chinese]. Academia Sinica, Institute of Zoology; Peking.

Each volume is devoted to individual families of insects. Some 16 volumes of this series have already been issued, and they still(!) continue to be published. Keys for identification are given. The books are well illustrated, including some colour plates.

Cornwell, P.B. *The cockroach – see* entries 889 and 890

831 Darlington, A. 1975. *The pocket encyclopaedia of plant galls in colour.* 191pp. Blandford Press; Poole, England.

This small volume is written in a semi-popular style. It is extremely well illustrated, making for easy identification of common galls in the UK.

832 Drew, R.I.A. *et al.* 1978. *Economic fruit flies of the South Pacific region.* 137pp. Queensland Department of Primary Industries; Brisbane.

A brief but useful account of this important pest. Three succinct chapters cover the subject: taxonomy, sterile insect release methods for suppression, and chemical methods for suppression.

833 Ebeling, W. 1959. *Subtropical fruit pests.* 436pp. Division of Agricultural Sciences, University of California; Berkeley, California.

Although the section on control methods and materials is now rather dated, those dealing with the biology of the pests themselves remain extremely useful. There are many line drawings and black-and-white photographs, together with a number of colour plates. Keys are provided, together with an extensive bibliography and a good index to scientific and common names.

834   Edwards, R. 1980. *Social wasps: their biology and control.* 398pp. Rentokil; East Grinstead, England.

A very readable, well illustrated and comprehensive work, aimed at general readers, food manufacturers, environmental health officers, pest controllers and all students of wasps. Pictorial keys are provided for the identification of the British species, and distribution in the British Isles is shown with maps. A useful checklist is included, giving the scientific names of the world species. Six hundred references are included in the bibliography, which also serves as the author index.

835   Edwards, C.A. and **Heath, G.W.** 1964. *The principles of agricultural entomology.* 418pp. Chapman & Hall; London.

Written for agricultural students and advisory officers, and for interested farmers with pest problems. Highly technical terms have been avoided as far as possible, but where used these are fully explained. Sections include: general principles of entomology and pest control; descriptions, bionomics and control of pests; and keys to the major pests. The text is well illustrated by black-and-white photographs and line drawings, and there is a useful bibliography.

836   Entwistle, P.F. 1972. *Pests of cocoa.* 779pp. Longman; London.

The standard work on this subject, including sections on the cocoa tree, its biology and fauna; pests arranged by orders; insects in relation to pollination; storage pests; a field guide to pest identification; and a systematic list of the pests showing their distribution. Although the many references are scattered through the book, there is a good author index.

837   Feltwell, J. 1981. *The large white butterfly: the biology, biochemistry and physiology of* Pieris brassicae (*L.*). 542pp. Dr W. Junk; The Hague. (*Series Entomologica* No. 18.)

'A source book for relevant literature and an introduction to all aspects of the white butterfly.' Some 41 000 references to key papers on the insect are included.

838   Fennemore, P.C. 1981. *Applied entomology: an introduction.* 200pp. Butterworths of New Zealand; Wellington.

Aims to provide a basic introduction to insects as living organisms and to the principles and practice of pest control. The treatment is introductory rather than exhaustive, and is not so heavily New Zealand-biased as to limit the usefulness of the book elsewhere. Includes a glossary and a list of scientific and common names.

839   Frankie, G.W. and **Koehler, C.S.** (Eds). 1978. *Perspectives in urban entomology.* 417pp. Academic Press; New York and London.

Deals with various aspects of insects, including their management, in urban environments. Strong American bias, with all but one chapter by American authors.

840 **Ferro, D.N.** (Ed.). 1976. *New Zealand insect pests.* 311pp. Lincoln University College of Agriculture; Canterbury, New Zealand.

Really the only up-to-date work on this topic, well written and illustrated with black-and-white photographs and line drawings. Pests are grouped according to the crop or activity of man that is affected, and cross-referenced where necessary. A brief description and discussion of the biology and control of each pest is presented and, although specific pesticides are not generally mentioned in the text, there is a separate chapter on insecticides. There is a good index to scientific and common names.

841 **Fröhlich, G.** and **Rodewald, W.** 1970. *Enfermedades y plagas de las plantas tropicales: descripcion y lucha.* 376pp. Edition Leipzig; Leipzig.

The work is arranged by crops with descriptions of their pests. Few of the plates are in colour, but they show the type of damage in addition to the insects. Each chapter has a short, not very full, bibliography.

842 **Grist, D.H.** and **Lever, R.J.A.W.** 1969. *Pests of rice.* 520pp. Longman; London.

The standard work in this field, covering the biology and control of all the major and most of the minor pests of rice. There is an extensive bibliography and a separate index to scientific names showing synonyms and misidentifications.

843 **Harris, K.F.** and **Maramorosch, K.** (Eds). 1977. *Aphids as virus vectors.* 559pp. Academic Press; New York and London.

A very full and comprehensive treatment of this economically important subject. Complementary to entry 844.

844 **Harris, K.F.** and **Maramorosch, K.** (Eds). 1980. *Vectors of plant pathogens.* 467pp. Academic Press; New York and London.

All the major vectors of plant pathogens are included in this authoritative work, which is intended for use by students, teachers and researchers of vector–pathogen–plant relationships. Aphids, leafhoppers and planthoppers, although mentioned briefly in this volume, are dealt with in more detail in Harris and Maramorosch (1977) and Maramorosch and Harris (1979), (entries 843 and 863).

Harris, W.V. *Termites: their recognition and control – see* entry 891

Hickin, N.E. *Termites: a world problem* and *The woodworm problem – see* entries 893 and 894

845 **Hill, D.S.** 1975. *Agricultural insect pests of the tropics and their control.* 516pp. Cambridge University Press; London.

A useful and well-arranged text despite a number of taxonomic errors. There are sections on the principles and methods of pest control; biological control in Africa; chemical control; pest descriptions, biology and control measures; and major crops and their pests. For each pest, information is given on its major and alternative hosts, its pest status, life history and distribution. Includes an extensive bibliography, a glossary and a good index to scientific and common names.

846   Hope, F. *et al.* 1969. *Recognition and control of pests and diseases of farm crops.* 166pp. Blandford Press; Poole, Dorset.

A most useful set of colour plates showing damage caused by insect pests and in many cases the insect (adult and larvae) causing the damage. There are indexes of English and Latin names. The text is rather limited, but the plates are the strength of this particular volume.

847   Hussey, N.W., Read, W.H. and **Hesling, J.J.** 1969. *The pests of protected cultivation: the biology and control of glasshouse and mushroom pests.* 404pp. Edward Arnold; London.

A very useful and comprehensive guide to the identification, biology, ecology, and control of pests of these crops. Reference is made to biological control methods where applicable, and there are numerous photographs both of the pests and of the damage they cause.

848   International Rice Research Institute. 1967. *The major insect pests of the rice plant.* 729pp. Johns Hopkins Press; Baltimore, Maryland.

A collection of papers given at a symposium of the International Rice Research Institute (1964). The sections cover taxonomy, population dynamics, the status of work on various pests, and losses and chemical control of the rice stem borers.

849   Johnson, W.T. and **Lyon, H.H.** 1976. *Insects that feed on trees and shrubs.* 463pp. Cornell University Press; Ithaca, New York.

Some 650 insect pests are described, together with the damage that they cause. The volume is lavishly illustrated with 212 coloured plates. There is a good selection of references for further reading, and good indexes, including an index to insects and host plants.

850   Jones, F.G.W. and **Jones, M.G.** 1974. *Pests of field crops.* 2nd edn. 448pp. Edward Arnold; London.

About half of this book is given over to a general introduction to insect structure and classification, and to descriptions of the biology of the various insect pests affecting field crops. There are also sections on pests of stored grain, and on pest management. There are many good line drawings and black-and-white photographs throughout, and an extensive bibliography.

851   Kalshoven, L.G.E. and **Devecht, J. van.** 1950. *De plagen van de cultuurgewassen in Indonesië.* 2 volumes. W. van Hoeve; The Hague.

This is a standard text for the region, and has not really been superseded. An English translation was produced by the publishers for the text for the illustrations of the work. The illustrations are on the whole rather poor. Mammals and birds are also included. A full English translation is now available (entry 852).

852 Kalshoven, L.G.E. 1981. *Pests of crops in Indonesia.* Revised and translated by P.A. van der Laan. 701pp. P.T. Ichtiar Baru. Van Hoeve; Djakarta.

A much-needed English edition of entry 851. Most of the original illustrations have been retained and some new ones added; many of the plates are in colour. There is an extensive bibliography, a species index and a useful host plant/pest index which includes Latin/English/Malayan crop names. An indispensable work for this region.

853 Kitching, R.L. and **Jones, R.E.** 1981. *The ecology of pests: some Australian case histories.* 254pp. CSIRO; Melbourne.

Eight of the 12 case histories presented here are entomological including, for example, the sheep blowfly and the *Sirex* woodwasp. Each species is dealt with in surprising depth considering the limitations of space, and several pages of references are given for each.

854 Knight, F.B. and **Heikkenen, H.J.** 1980. *Principles of forest entomology.* 5th edn. 461pp. McGraw-Hill; New York.

A very comprehensive work, covering all aspects of the science and practice of forest entomology, including not only pests of forest trees and their products, but also forest arthropods of medical importance. There is a list of scientific and common names, arranged systematically, a short glossary and a good general index that includes both scientific and common names.

855 Kono, T. and **Papp, C.S.** 1977. *Handbook of agricultural pests: aphids, thrips, mites, snails and slugs.* 205pp. Department of Food and Agriculture; Sacramento, California.

A useful little book based on an identification manual devised for training survey biologists. The majority of the species treated in the keys and figures have been widely distributed by man through commerce, so the handbook is of international use.

856 Kranz, J., Schmutterer, H. and **Koch, W.** (Eds). 1977. *Diseases, pests and weeds in tropical crops.* 666pp. Verlag Paul Parey; Berlin and Hamburg.

An unparalleled work, containing contributions by 152 authors, providing information on the distribution, biology, control and economic importance of all the more important pests of tropical crops. Well illustrated throughout with line drawings and plates, including some superb colour photographs. Pests are indexed under host plants and also by their scientific and common names.

Krishna, K. and **Weesner, F.M.** *Biology of termites – see* entry 897

857   Lamb, K.P. 1974. *Economic entomology in the tropics.* 195pp. Academic Press; London and New York.

Orientated towards the needs of developing countries in the humid tropics, emphasis being placed on the agricultural pests of the Oriental and Australian regions. However, the major pests of crops from other parts of the tropics are also considered, together with insects of medical and veterinary importance. Information on the biology and ecology of the various groups is systematically arranged, and there is a short bibliography at the end of each section. There are a number of inconsistencies between the text and the index to scientific and common names, and a number of actual errors in each. Nevertheless, this is a useful, but by no means comprehensive, introductory text.

858   Lavabre, E.M. 1970. *Insectes nuisibles des cultures tropicales (Cacaoyer, Caféier, Colatier, Poivrier, Théier).* 276pp. G.-P. Maisonneuve & Lavose; Paris.

The first four chapters describe methods of control, mostly by insecticides, and are now mostly outdated by legislation concerning the use of certain of these insecticides. The remaining chapters deal with the pests. The book is well illustrated with line drawings. Good descriptions of the pests are given along with the damage they cause. There is a useful but somewhat out-of-date bibliography.

859   Le Pelley, R.H. 1968. *Pests of coffee.* 590pp. Longman; London.

The standard work on this subject. Provides detailed information on the biology, ecology and control of all types of coffee pest, arranged systematically. There is a useful checklist of recorded coffee insects and an extensive bibliography.

860   McCrae, D.J. 1981. *Insects of agricultural importance in Brunei.* 110pp. Department of Agriculture; Bandae Seri Begawan, Brunei.

The only up-to-date work for this area, providing lists of harmful and beneficial species indicating their hosts. Well indexed.

861   McKelvey, J.J., Jr, Eldridge, B.F. and **Maramorosch, K.** (Eds). 1981. *Vectors of disease agents: interactions with plants, animals and man.* 243pp. Praeger Scientific; New York.

The major objective of this comprehensive volume is to present the state of the art and the science of vector–pathogen–host relationships. Contains a useful glossary of specialized terminology, an extensive bibliography and a good index to authors, species and subjects.

862   Mani, M.S. 1974. *Ecology of plant galls.* 434pp. Dr W. Junk; The Hague. (*Monographiae Biologicae* No. 12.)

An outline of the more outstanding features of the general ecology of plant galls. The insects causing and associated with the galls are described in some detail. There is a very full bibliography; it is now somewhat outdated, but some most useful references are included.

863 Maramorosch, K. and **Harris, K.F.** (Eds). 1979. *Leafhopper vectors and plant disease agents.* 654pp. Academic Press; London.

Another important work in the same series as, and complementary to, *Vectors of plant pathology* by Harris and Maramorosch (entry 844).

864 Metcalf, C.L. and **Flint, W.P.** 1951. *Destructive and useful insects: their habits and control.* 3rd edn. McGraw-Hill; New York.

Although parts of this work are now out of date, particularly the control section, a mass of useful information is included. This is an often quoted work, and is heavily used as a textbook for economic entomology and as a reference book of its principles and practices. The work has a good index.

865 Nickel, J.L. 1979. *Annotated list of insects and mites associated with crops in Cambodia.* 75pp. Southeast Asian Regional Center for Graduate Study and Research in Agriculture; College, Laguna, Philippines.

Although the field observations, collections and photographs for this work were only about half completed when the USAID personnel were forced to leave Cambodia in 1963, it is a unique and therefore valuable contribution for this area. Pests are arranged according to host plant, and there is a short section on stored-products pests. There are 22 black-and-white photographs and a useful checklist of species.

866 Ordish, G. 1976. *The constant pest: a short history of pests and their control.* 240pp. Peter Davies; London.

A good general introduction to this interesting subject. Includes a useful bibliography and a brief glossary.

867 Pearson, E.O. 1958. *The insect pests of cotton in tropical Africa.* 355pp. Empire Cotton Growing Corporation and the Commonwealth Institute of Entomology; London.

The book is concerned with the insects that are associated with cotton in Africa south of the Sahara. The main portion of the book consists of an account of the major pests of the crop. It covers the taxonomy and distribution of the pests, their life histories and seasonal activity. Minor pests are included, but not covered in the fuller style allocated to those of a more important nature. There is a bibliography of some 686 references. Eight colour plates are included showing the various stages of the insect pests and the damage they can cause.

868 Pedgley, D. 1982. *Wind-borne pests and diseases: meteorology of air-borne organisms.* 250pp. Elliswood; Chichester.

The book describes and explains the influence of the atmosphere on the wind-borne movement of small organisms. As many pests of agriculture and medically important insects are carried by wind currents, etc., this is a most useful contribution to the literature of this subject. There are some 40 pages of references and the author has obviously kept these up to date whilst the book was in press. A good subject index, but perhaps not as full as it might be.

869   Pfadt, R.E. (Ed.). 1978. *Fundamentals of applied entomology.* 3rd edn. 798pp. Macmillan; New York.

A most useful and comprehensive text which includes a chapters on all the major aspects of applied entomology, with sections on the pests of various crops or groups of crops, and on insects of medical and veterinary importance. Includes a useful glossary.

870   Pirone, P.P. 1978. *Diseases and pests of ornamental plants.* 5th edn. 566pp. John Wiley; New York.

First published in 1943, this official publication of the New York Botanical Garden is a guide to the diagnosis and treatment of diseases and organisms afflicting nearly five hundred genera of ornamental plants grown outdoors, under glass or in the home. Numerous black-and-white photographs throughout, and a good index to scientific and common names.

871   Pruthi, H.S. 1969. *Textbook on agricultural entomology.* 977pp. Indian Council of Agricultural Research; New Delhi.

A very comprehensive work, providing a good general introduction to entomology, including collecting, rearing and preservation techniques, a section on the history of entomology in India, and much useful information on the more important harmful and beneficial insects in India. There are numerous line drawings and plates and a useful bibliography arranged under broad subject headings. Keys are included as an appendix and the whole work is well indexed.

872   Rajpreecha, J. 1980. *Insect pests of cashew in Thailand.* [In Thai]. 182pp. National. Biological Control Research Center, Kasetsart University; Bangkok. (Special Report No. 2.)

Major insect pests species are described including their biology, distribution and the extent of the damage caused. The work is written in Thai, but there is an English summary.

873   Schwenke, W. (Ed.). 1972–. *Die Forstschadlinge Europas.* 5 volumes. Verlag Paul Parey; Hamburg and Berlin.

A very detailed and comprehensive work updating Escherich's *Die Forstinsekten Mitteleuropas.* Intended as a manual rather than a textbook, the first four volumes of Schwenke's work provide detailed information on species that are forest pests in Europe (Volume 5 deals with vertebrates). Illustrated by line drawings and well indexed.

874   Shuidao, H.T.T. 1978. *Illustrations of rice and their natural enemies.* [In Chinese]. 179pp. People's Press; Hubei.

Eighty-eight colour plates of pests; some show the area of the plant normally damaged by the pest, and areas where eggs are laid. There is an index of Latin names with Chinese equivalents.

875   Simmonds, N.W. 1966. *Bananas.* 2nd edn. 512pp. Longman; London.

Includes a useful annotated list of banana pests, systematically arranged, and detailed information on the biology, distribution, effects and control of the major pests.

876   Singh, S.R., Van Emden, H.F. and **Taylor, T.A.** (Eds). 1978. *Pests of grain legumes: ecology and control.* 454pp. Academic Press; London and New York.

Comprises the proceedings of an International Symposium on Pests of Grain Legumes held in Nigeria in 1976, and provides a very comprehensive treatment of this important subject. There is a 42-page bibliography, and a useful systematic checklist of insect and mite pests of grain legumes that gives common names where appropriate.

877   Smit, B. 1964. *Insects in Southern Africa and how to control them.* Oxford University Press; Cape Town.

A useful work for this area. The major pests, their distribution and life histories are described. The work has a good index that includes Afrikaans names for the insects described.

878   Spencer, K.A. 1973. *Agromyzidae (Diptera) of economic importance.* 418pp. Dr W. Junk; The Hague, Netherlands.

The standard work on this important family, providing keys for identification, detailed information on biology, ecology and control, and an extensive bibliography.

879   Swain, G. 1971. *Agricultural zoology in Fiji.* 424pp. HMSO; London.

Written by a former Fiji government entomologist, this book presents knowledge gained concerning pests and beneficial animals in Fiji up until the late 1960s. It is almost exclusively entomological in content, and places strong emphasis on identification and control of pest species. Includes nearly four hundred photographs and an extensive bibliography.

880   Tazima, Y. (Ed.). 1978. *The silkworm: an important laboratory tool.* 307pp. Kodansha; Tokyo.

A comprehensive review of all aspects of the commercial silkworm, *Bombyx mori*, including its rearing and laboratory uses.

881   Treat, A.E. 1975. *Mites of moths and butterflies.* 362pp. Cornell University Press; Ithaca (New York) and London.

Although peripheral to mainstream entomology, this book is included here as it is the only comprehensive work on mite–lepidopteran associations.

882   Vasantharaj David, B. and **Kumaraswami, T.** 1975. *Elements of economic entomology.* 508pp. Popular Book Depot; Madras, India.

A very comprehensive work, including sections on the history of entomology in India; forecasting; pests (arranged by hosts); beneficial insects; integrated control; pesticides and pesticide application; non-insect pests; etc. There is a good index to scientific and common names.

883 Williams, J.R., Metcalfe, J.R., Mungomery, R.W. and **Mathes, R.** (Eds). 1939. *Pests of sugar cane.* 568pp. Elsevier; Amsterdam.

The standard work in this field, presenting general information on sugar cane pests, their distribution, taxonomy and control, and on crop loss assessment methods, together with individual sections on specific pests. Each chapter includes an extensive bibliography, although there is no overall author index.

884 Wolcott, G.N. 1933. *An economic entomology of the West Indies.* 688pp. Entomological Society of Puerto Rico; San Juan.

Although now rather dated, this is probably the only comprehensive work for this area. There is a general introduction to entomology, a section on ecology and one on control. Pests are grouped according to the type of crop they attack, and many of the species mentioned are illustrated by line drawings. There are some 174 references scattered through the text, and these are included in the very full author/subject index.

885 Wyniger, R. 1962 Pests of crops in warm climates and their control. *Acta Tropica Suppl.* **7**: 1–555.

The work is issued in two parts. Part 1 is arranged by crop. The damage caused is described and a brief description of the pest involved is given along with its distribution. Control measures are described in Part 2, but this is now somewhat out of date.

886 Yunus, A. and **Hua, H. T.** 1980. List of economic pests, host plants, parasites and predators in West Malaysia 1920–1978. *Bulletin, Ministry of Agriculture, Malaysia* No. **153**: 1–538.

Although this list does include animals other than insects, some 90% of the pest names are insects.

INSECTS AS HOUSEHOLD AND INDUSTRIAL PESTS

887 Bell, W.J. and **Adiyodi, K.G.** (Eds). 1981. *The American cockroach.* 529pp. Chapman & Hall; London.

An integrated account of the biology of the American cockroach. A pest of stored products, the insect is also used widely as a laboratory animal. The volume gives an all-round coverage of all aspects of the biology of this insect. The illustrations are rather poor, but there is a good and full bibliography.

888 British Museum (Natural History). Economic Series. No. 11a (1979), *Domestic wood-boring insects* (40pp); No. 15 (1980), *Common insect pests of stored food products.*

A useful series of pamphlets. They are well illustrated and in some cases give keys for identification. Suggestions for control are offered.

889 Cornwell, P.B. 1968. *The cockroach*; Vol. 1, *A laboratory insect and an industrial pest.* 391pp. Hutchinson; London. (The Rentokil Library.)

890 Cornwell, P.B. 1976. *The cockroach*; Vol. 2, *Insecticides and cockroach control.* 557pp. Associated Business Programmes; London. (The Rentokil Library.)

Although these two volumes are designed to be used separately if required, they combine to provide a comprehensive and easy-to-read text on these important insects. Volume 1 gives an account of the biology of the more common species, including details of their structure, physiology, behaviour and ecology. The references in this volume are, unfortunately, arranged numerically rather than alphabetically. Volume 2 details the insecticides, formulations and equipment used for cockroach control, together with information on safety, resistance and test procedures. The extensive bibliography also serves as an author index to this volume. Both volumes are well illustrated by photographs and line drawings.

891 Harris, W.V. 1971. *Termites: their recognition and control.* 7th edn. 186pp. Longman; London.

A standard work, covering virtually every aspect of this subject, on a worldwide basis. Well illustrated by line drawings and many photographs, some in colour.

892 Hickin, N.E. 1964. *Household insect pests: an outline of the identification, biology and control of the common insect pests found in the home.* 172pp. Hutchinson; London. (The Rentokil Library.)

A comprehensive and well illustrated work.

893 Hickin, N.E. 1971. *Termites: a world problem.* 232pp. Hutchinson; London. (The Rentokil Library.)

An authoritative survey of the biology, economic significance and methods of control of termites in both natural and man-made environments. A set of three wall charts included in a pocket show, among other things, the anatomy of termites and their classification, with illustrations to distinguish the families and most of the subfamilies. The text is well illustrated throughout.

894 Hickin, N.E. 1972. *The woodworm problem.* 2nd edn. 123pp. Hutchinson; London. (The Rentokil Library.)

A comprehensive guide to the identification, biology and control of the various wood-boring beetles known collectively as woodworm. Also includes a considerable amount of information on other insects commonly mistaken for these pests, and on other insects that contribute to wood decay. Well illustrated.

895 Hickin, N.E. 1975. *The insect factor in wood decay: an account of wood-boring insects with particular reference to timber indoors.* 3rd edn, revised and edited by R. Edwards. 383pp. Associated Business Programmes; London. (The Rentokil Library.)

Gives a comprehensive account of the insects that damage wood in buildings in Britain. Extremely well illustrated by 280 drawings and photographs.

896 Hinton, H.E. 1945. *A monograph of the beetles associated with stored products*, Vol. 1. 443pp. British Museum (Natural History); London.

Still the standard work on this subject, providing keys for the identification of 200 species, of which 175 are fully described in the text. Includes an extensive bibliography. Volume 2 of this work was never actually published.

897 Krishna, K. and **Weesner, F.M.** 1969. *Biology of termites.* 2 volumes. Academic Press; London.

This very detailed and authoritative work deals with the fundamental aspects of termite biology. The areas considered include embryology, nesting, taxonomy and the termite faunas of the various zoogeographical regions. Each chapter has its own full bibliography, and subject indexes are included in each volume.

898 Munro, J.W. 1966. *Pests of stored products.* 234pp. Hutchinson; London. (The Rentokil Library.)

A general account of insects that infest products in storage, and the measures available for their prevention and control. Well illustrated.

PEST CONTROL

899 Brown, A.W.A. 1978. *Ecology of pesticides.* 525pp. John Wiley; New York and Chichester.

A detailed review of the research work performed on this controversial subject. Good indexes to authors, subjects and species names.

900 Burges, H.D. (Ed.). 1981. *Microbial control of pests and plant diseases 1970–1980.* 949pp. Academic Press; London and New York.

Continues from, and complements entry 901.

901 Burges, H.D. and **Hussey, N.W.** (Eds). 1971. *Microbial control of insects and mites.* 861pp. Academic Press; London and New York.

An extremely comprehensive review of the subject, including useful appendices on information sources and literature searching in biological control, information services, sources of special materials, and a bibliography on diseases and enemies of medically important arthropods from 1963 to 1967. Includes a useful author index to the numerous references in the text. Now updated and complemented by entry 900.

902 Centre for Overseas Pest Research. PANS Manuals: No. 1 (1971), *Pest control in bananas* (128pp); No. 2 (1973), *Pest control in groundnuts* (197pp); No. 3 (1970), *Pest control in rice* (270pp); No. 4 (1978), *Pest control in tropical root crops* (235pp); No. 5 (1981), *Pest control in tropical grain legumes* (206pp).

This series deals with individual crops, each manual covering in some detail all the insects that are pests of the particular crop, including the life cycle and distribution. In most cases common names are included.

903 Clausen, C.P. 1940. *Entomophagous insects.* 688pp. McGraw-Hill; New York.

Still regarded as the standard work on this subject, this book represents what the author would have liked to have had available while engaged in fieldwork himself. The main text is systematically arranged and very fully indexed, and there is an extensive bibliography.

904 Coppell, H.C. and **Mertins, J.W.** 1977. *Biological insect pest suppression.* 314pp. Springer-Verlag; Berlin, Heidelberg and New York.

Defines biological insect pest suppression as 'the use or encouragement, by man, of living organisms or their products for the population reduction of pest species'. A very comprehensive work, including chapters on the history of biological control, the organisms used, and integrated pest control. There is a useful glossary and an extensive bibliography.

905 Cantwell, G.E. (Ed.). 1974. *Insect diseases.* 2 volumes. 595pp. Marcel Dekker; New York.

Provides detailed information on the wide range of diseases affecting insects, and the ways in which these can be manipulated for pest control. Diagnostic techniques are described, and there are laboratory exercises accompanying each chapter. A useful glossary is included, and there are good indexes to authors and species. Illustrated with many photographs, and contains some nine hundred references.

906 DeBach, P. 1974. *Biological control by natural enemies.* 323pp. Cambridge University Press; London.

Outlines the workings and potentials of biological control for the general reader and student interested in environmental phenomena. There are chapters tracing the history of biological control and, in others, case histories of control programmes are given. An excellent introduction to a subject of growing importance.

907 Flint, M.L. and **Van den Bosch, R.** 1981. *Introduction to integrated pest management.* 240pp. Plenum; New York.

Presents a comprehensive review of the basic principles and methods of integrated pest management, assuming minimal previous knowledge on the part of the reader. Includes a number of case histories and a quite lengthy bibliography.

908 Greathead, D.J. (Ed.). 1971. *A review of biological control in the Ethiopian region.* 162pp. Commonwealth Agricultural Bureaux; Farnham Royal. (*Technical Communication, Commonwealth Institute of Biological Control* No. 5.)

Reviews biological control within the Ethiopian biogeographical region up to mid-1969. There is a very extensive bibliography and an index to scientific and, in some cases, common names.

909 Greathead, D.J. (Ed.). 1976. *A review of biological control in western and southern Europe.* 182pp. Commonwealth Agricultural Bureaux; Farnham Royal. (*Technical Communication, Commonwealth Institute of Biological Control* No. 7.)

Reviews biological control from the first successful introduction of a natural enemy into Europe in 1897, up to 1974–5. There is a very extensive bibliography and an index to scientific and some common names.

910   Gunn, D.L. and **Rainey, R.C.** 1979. Strategy and tactics of control of migrant pests. *Philosophical Transactions of the Royal Society of London, B* **287**: 249–488.

A discussion meeting at the Royal Society of London. The coverage includes locust control, *Simulium* and Onchocerciasis, African armyworm and the migratory grasshopper.

911   Hill, D.S. and **Waller, J.M.** 1982. *Pests and diseases of tropical crops;* Vol. 1, *Principles and methods of control.* 175pp. Longman; London.

An excellent book, designed for agriculture students and agricultural extension workers. Well-written and very thorough, illustrated throughout with black-and-white photographs and line drawings, and including a useful bibliography. Volume 2 will give details of the individual pests and diseases and the crops they attack.

912   Huffaker, C.B. and **Messenger, P.S.** (Eds). 1976. *Theory and practice of biological control.* 788pp. Academic Press; New York and London.

Very much the standard work in its field, this volume provides a useful background to the history, theory and application of biological control. One chapter is devoted to a systematically arranged table of pests, giving details of biological programmes which have been used against them. The text is well supported by references, but unfortunately these are not collected into a single bibliography and are not indexed.

913   Jacobson, M. 1972. *Insect sex pheromones.* 382pp. Academic Press; New York and London.

Perhaps now slightly out of date, this is nevertheless a very useful and comprehensive work, covering all aspects of this subject. A large part of the book is given over to a bibliography of some 1400 references.

914   Kalmakoff, J. and **Longworth, J.F.** 1980. *Microbial control of insect pests.* 102pp. Department of Scientific and Industrial Research; Wellington, New Zealand. (*DSIR Bulletin* No. 228.)

Compiled from lecture and practical notes presented at a Unesco/UNEP/ICRO regional training course held in New Zealand in 1977. The emphasis is strongly on practical application throughout.

915   Metcalfe, R.L. and **Luckmann, W.H.** (Eds). 1975. *Introduction to insect pest management.* 587pp. John Wiley; New York.

Aims to provide an introduction to the subject and to stimulate the adoption and use of pest management systems. Comprehensive and clearly written with plenty of examples and an extensive bibliography.

916   Pimentel, D. (Ed.). 1981. *CRC handbook of pest management in agriculture.* 3 volumes. CRC Press; Boca Raton, Florida.

A monumental set of volumes presenting quite the most up-to-date and comprehensive guide to this very large and complex subject. Many of the authors are well-known world authorities in their particular subject fields, and much of the information given is unavailable elsewhere. Each volume can be used on its own, and has its own subject index. The only criticism might be that there is no index to the very large number of references embodied in the text.

917 Poinar, G.O., Jr and **Thomas, G.M.** 1978. *Diagnostic manual for the identification of insect pathogens.* 218pp. Plenum; New York and London.

Provides illustrated keys for the identification of pathogens causing infectious diseases of insects. Although the emphasis is on identification, general background information on the various groups of pathogens is also presented. An extensive bibliography completes this most useful work.

918 Scopes, N. and **Ledieu, M.** (Eds). 1979. *Pest and disease control handbook.* 450pp. British Crop Protection Council; Croydon, England.

This is essentially an updating and revision of the 5th edition of the well-known *Insecticide and fungicide handbook*, and provides up-to-date information on the control of pests and diseases, excluding those of forest crops and those of outdoor ornamental plants.

919 Van den Bosch, R., Messenger, P.S. and **Gutierrez, A.P.** 1982. *An introduction to biological control.* 230pp. Plenum; London and New York.

A thorough update of Van den Bosch and Messenger's earlier *Biological control.* Includes current research findings and new information on insect pathology, population dynamics, integrated pest management, and the economics of biological control. Suitable for use by undergraduate students.

920 Ware, G.W. 1980. *Complete guide to pest control – with and without chemicals.* 290pp. Thomson Publications; Fresno, California.

Aims to meet the needs of urban and suburban residents, describing the pest problems that might be encountered anywhere in the USA, and suggesting solutions. Despite the emphasis on the USA, the information given would in many cases be applicable to other countries.

5 Searching and locating the literature

SECONDARY JOURNALS

Elsewhere in this guide we have listed a mere selection of those primary or research journals which may be regarded as central to the subject of entomology. There are many others which, for one reason or another, we have not included but which nevertheless deal exclusively with entomology, and a still greater number which occasionally include entomological papers.

We estimate that there are perhaps some six hundred primary journals that are wholly devoted to entomology and which therefore constitute the core of the literature, and another thousand or so fringe journals which, regularly or irregularly, cover some aspect of the subject. Authors are often faced with the dilemma of publishing either in one of the core journals, which may entail a lengthy delay, or in one of the fringe journals that can offer fairly rapid publication. As speed of publication is often of considerable importance, particularly to authors of taxonomic papers, an increasing number of important papers are now appearing in the fringe literature. Within the field of applied entomology, the problems posed by such 'scatter' are even greater, largely because the subject is by nature interdisciplinary, and significant papers appear in the journals of plant pathology, medicine, veterinary science, chemistry, etc.

Of course, journals are not the only outlet for research papers, and anyone wishing to keep fully up to date also needs to be aware of the many books, conference proceedings, reports and theses which are published each year. In many ways, these are even more difficult to identify than journals, as they come from a seemingly unlimited variety of sources, are often poorly advertised and may be produced only in small numbers for limited distribution.

It is clear that researchers need some kind of key to this wealth of literature if they wish to maintain anything more than a broad overview of the subject, or wish to carry out reasonably exhaustive literature searches. This is where the various secondary journals can be of enormous but widely underestimated value, specializing as they do in identifying, locating, and abstracting or indexing all of the literature within their defined fields and limits.

When one is selecting a secondary journal for any specific purpose, it is important to remember that they do vary considerably in scope and emphasis and although their coverage may overlap, there are invariably unique references to be found in each. For this reason, it is usually worth consulting two or more such journals if an exhaustive literature search is required, although of course this may not always be practical.

There is much to be said for using one of the secondary journals that include abstracts, as these make it possible to evaluate the relevance of the references before possibly going to the expense and trouble of obtaining a copy of the original paper. It is generally recognized that the titles of scientific papers are relatively poor indicators of their actual content or, worse still, may actually be positively misleading. It follows that a selection made using one of the 'title-only' secondary journals will, on the whole, be much less useful than one made using an abstracting journal proper.

Even within the realm of abstracting journals there are distinct differences which should be borne in mind. Some journals of this type provide only a brief indicative abstract, giving an outline of what the paper is about, whereas others give a fairly detailed informative abstract, which includes much or all of the scientific data from the original paper. While the former type are useful in assessing the relevance of a paper, the latter can sometimes actually stand in for the original text. This is particularly useful where the original paper is in a 'difficult' language, as virtually all of the abstracting journals are published in English, and therefore provide the non-linguist with a kind of mini translation service.

As with primary journals, secondary journals may be divided into those which are central to the subject and those which cover only certain aspects of it. We have included a selection of the latter in the following list, as it is not always immediately obvious that they are relevant to entomology and, for that reason, they are frequently overlooked.

Many of the secondary journals listed here may be speedily and easily searched using one of the publicly available 'online' computer systems, and this aspect of their use is discussed on p. 155 *et seq*.

921   Abstracts of Entomology. 1970–. 12 per year. En. BioSciences Information Service (BIOSIS), 2100 Arch Street, Philadelphia, Pennsylvania 19103, USA.

Comprises a mixture of informative abstracts and 'title-only' bibliographic citations of pure and applied research publications covering insects, arachnids and insecticides. Derived from *Biological Abstracts* and *Bioresearch Index*. Although the coverage is good, the contents are not systematically arranged and the indexes are complicated and difficult to use. The annual cumulative index is not included in the basic subscription.

922   Abstracts on Hygiene and Communicable Diseases. 1968–. 12 per year. En. Bureau of Hygiene and Tropical Diseases, Keppel Street, London WC1E 7HT, England.

Includes abstracts relevant to medical entomology. Monthly author and subject indexes with annual cumulations included in the basic subscription.

Acridological Abstracts – *see* entry 348

923 Agrindex. 1975–. 12 per year. En. Apimondia, Str. Pitar Mos No. 2, Bucureşti, Romania.

Comprises bibliographic citations and a limited number of abstracts, supplied by regional input centres, covering the literature of agricultural sciences and technology. Part of the input is supplied by the CAB, so there is an element of overlap with, for example, the *Review of Applied Entomology*. The main weakness of *Agrindex* lies in its limited number of actual abstracts; also, for the serious researcher, its emphasis on extension literature can present problems. Without an abstract to make the level of a paper clear, it is all too easy to mistake something quite trivial for a serious research paper.

924 Apicultural Abstracts. 1950–. 4 per year. En. International Bee Research Association, Hill House, Gerrards Cross, Buckinghamshire SL9 0NR, England.

An abstracting journal proper, covering the world literature on bees, including their pollinating activities and the rearing of *Apis* and other Apoidea for pollination. Separate annual author and subject indexes.

925 Bibliographie der Pflanzenschutzliteratur. 1914–. 4 per year. En. Verlag Paul Parey (Berlin), Lindenstr. 44–47, D-1000 Berlin 61, Federal Republic of Germany.

Comprises bibliographic references only, covering all aspects of plant protection. An English-language title is provided for each reference, and subject headings are given in English, German and French. Good author and subject indexes.

926 Bibliography of Agriculture. 1942–. 12 per year. En. Oryx Press, Suite 106, 2214 North Central at Encanto, Phoenix, Arizona 85004, USA.

Comprises bibliographic citations only, covering the world literature on agriculture and related subjects. Prepared by the National Agricultural Library (Beltsville, USA), *Bibliography of Agriculture* is effectively the library catalogue of NAL, and includes not only scientific and extension literature, but also a fairly high proportion of more general material of the kind found in any large library. References are arranged under broad but useful headings in each part, with a subject and author index. An annual cumulative index is produced, but is not included in the standard subscription. Because of its general nature, this publication has become rather bulky and perhaps a little cumbersome to use.

927 Bibliography of Indian Zoology. 1958–. 1 per year. En. Zoological Survey of India, 34 Chittanjan Avenue, Calcutta 700012, India.

Each volume includes: a brief review of the work published in India and elsewhere on the fauna of the Indian region; an extensive bibliography (bibliographic citations only) arranged in author order; a list of new taxa; and a systematic index. Publication has become both slow and rather erratic of late, with recent volumes appearing some 11 or 12 years late.

928   Bulletin Analytique d'Entomologie Médicale et Vétérinaire. 1954–. 12 per year. Fr. Service Publications ORSTOM, 70–74 route d'Aulnay, F-93140 Bondy, France.

Bibliographic citations only (no abstracts) covering all aspects of medical and veterinary entomology. Entries are arranged under useful subject headings, with cross-references where necessary, and each is marked with a code to denote the library in the Paris area at which the original was seen. Each issue includes indexes to authors, geographic regions, conferences and meetings, and reports. This service is remarkable for its completeness and up-to-date references. These criteria make it an indispensable service to medical entomologists.

929   Current Contents (various series). 1959–. 52 per year. En. Institute for Scientific Information, 3501 Market Street, University City Science Center, Philadelphia, Pennsylvania 19104, USA.

Rather the odd-man-out in this section, *Current Contents* is not really an abstracting or indexing journal. Its publishers simply gather together the contents pages of a fairly large number of primary research journals (plus those of a few monographic series and similar publications) and photographically reproduce them. On the whole, *Current Contents* is very up-to-date, so that it may be used as a quick alerting service. The two series that are of interest to entomologists are *Life Sciences* and *Agriculture, Biology and Environmental Sciences*; a list is available showing which journals are covered by each.

930   Dissertation Abstracts International, B (The Sciences and Engineering). 1969–. 12 per year. En. University Microfilms International, Ann Arbor, Michigan 48106, USA.

A monthly compilation of abstracts of doctoral dissertations submitted to the publishers by more than 430 cooperating institutions in the USA and Canada. In most cases copies of the dissertations are available for purchase from University Microfilms. The abstracts are arranged under useful subject headings, including one for entomology and several for relevant aspects of agriculture, such as animal pathology, plant pathology, veterinary science, forests and wildlife. There is a reasonably easy-to-use keyword title index and an author index to each part, but only the author index is cumulated annually.

931   Entomology Abstracts. 1969. 12 per year. En. Information Retrieval Inc., Suite 815, Fisk Building, 250 West 57th Street, New York, New York 10019, USA.

Each issue comprises approximately 750 indicative abstracts, covering all aspects of entomology. The abstracts are arranged under useful headings and there are good monthly and annual indexes. Generally liked by entomologists, perhaps partly because it is simple to use, well laid out and clearly printed.

932   Field Crop Abstracts. 1948–. 12 per year. En. Commonwealth Bureau of Pastures and Field Crops, Hurley, Maidenhead, Berkshire SL6 5LR, England.

One of the CAB abstracting journals, covering the world scientific literature on annual field crops, temperate and tropical. Mainly deals with the agronomy, botany and physiology of these crops, but also includes material on pests and diseases. Mainly informative abstracts, well indexed. Some degree of overlap with *Review of Applied Entomology, A*, and other CAB abstracting journals.

933 Forest Products Abstracts. 1978–. 12 per year. En. Commonwealth Forestry Bureau, South Parks Road, Oxford OX1 3RD, England.

One of the CAB abstracting journals, covering all aspects of forest products but excluding forestry relating to the living trees. Includes material on insect pests of timber. Mainly informative abstracts, well indexed. Some degree of overlap with *Review of Applied Entomology, A,* and other CAB abstracting journals.

934 Forestry Abstracts. 1939–. 12 per year. En. Commonwealth Forestry Bureau, South Parks Road, Oxford OX1 3RD, England.

Covers the world literature on all aspects of forestry (excluding forest products from 1978), including forest insects and other invertebrates. Mainly informative abstracts, well indexed. Some degree of overlap with *Review of Applied Entomology, Series A*, and other CAB abstracting journals.

935 Helminthological Abstracts, Series A (Animal Helminthology). 1932–. 12 per year. En. Commonwealth Institute of Parasitology, 395A Hatfield Road, St Albans, Hertfordshire AL4 0XU, England.

One of the CAB abstracting journals, covering the world literature on animal helminthology, especially in relation to veterinary, medical, agricultural and fisheries science. Includes material relevant to medical and veterinary entomology, but there is some overlap with the *Review of Applied Entomology, Series B,* and other CAB journals. Mainly informative abstracts, well indexed.

936 Helminthological Abstracts, Series B (Plant Nematology). 1932–. 4 per year. En. Commonwealth Institute of Parasitology, 395A Hatfield Road, St Albans, Hertfordshire AL4 0XU, England.

Another CAB abstracting journal, covering the world literature on plant, free-living and insect nematology and related subjects. Mainly informative abstracts, well indexed. Some overlap with other CAB journals.

937 Horticultural Abstracts. 1931–. 12 per year. En. Commonwealth Bureau of Horticulture and Plantation Crops, East Malling Research Station, Maidstone, Kent ME19 6BJ, England.

Another CAB abstracting journal, covering the world literature on horticultural and plantation crops, both temperate and tropical. Includes material on pests and diseases. Mainly informative abstracts, well indexed. Some overlap with other CAB journals.

938   Index-Catalogue of Medical and Veterinary Zoology. 1932–. Irregular. En. Animal Parasitology Institute, USDA, SEA-AR, BARC-East, Building 1180, Beltsville, Maryland 20705, USA.

An index to the world literature on animal parasites of animals, including man. Issued as a series of supplements, in parts providing for different approaches, e.g. Part 1 is an alphabetic author listing, Part 7 is a systematically arranged host listing. Full bibliographic details are provided for each reference, together with added keywords. A useful feature is the inclusion in Part 1 of codes to denote libraries at which the original documents may be seen. A useful tool for retrospective searches on medical and veterinary entomology.

939   Index Veterinarius. 1933–. 12 per year. En. Commonwealth Bureau of Animal Health, Central Veterinary Laboratory, New Haw, Weybridge, Surrey KT15 3NB, England.

Another CAB journal, providing an author and subject index to the world veterinary literature. Within each monthly part, the bibliographic references (no abstracts) are arranged under subject headings, and there is a good index to each part. Rather than a cumulative index, a complete cumulated annual edition is issued to replace the monthly parts, and this is included in the basic subscription. *Index Veterinarius* includes all of the material abstracted in the *Veterinary Bulletin*, together with various categories of papers that are not usually included there, such as conference papers, fringe literature, etc. A useful journal for veterinary entomologists, but overlapping with other CAB journals.

940   Referativnyi Zhurnal, Biologiya. 1953–. 12 per year. Ru. Mezhdunarodnaya Kniga, Moscow G-200, USSR.

The Soviet abstracting journal for the biological sciences, each issue being divided into several sections. Section E (Entomologiya) covers all aspects of pure and applied entomology, with informative abstracts arranged under useful subject headings; Section K (Zooparazitologiya) covers insects and Acari as vectors of disease pathogens of man and animals. All the abstracts are in Russian, although the coverage is international, with 50% or more of the literature cited being from outside the Eastern bloc. A possible drawback is that authors' addresses are not given for any of the papers cited. There are annual indexes, but individual parts carry only a contents guide to each section.

941   Review of Applied Entomology, Series A (Agricultural). 1913–. 12 per year. En. Commonwealth Institute of Entomology, 56 Queen's Gate, London SW7 5JR, England.

One of the CAB abstracting journals, covering the world literature on insect and other arthropod pests of cultivated plants, forest trees and stored products of animal and vegetable origin; beneficial arthropods such as parasites and predators are included. Material on apiculture and sericulture is generally excluded however. Almost all the bibliographic references are accompanied by an informative abstract, and there are good monthly and annual author and subject indexes. There is some overlap with other CAB journals.

942   Review of Applied Entomology, Series B (Medical and Veterinary).
1913–. 12 per year. En. Commonwealth Institute of Entomology, 56
Queen's Gate, London SW7 5JR, England.

One of the CAB abstracting journals, covering the world literature on insects
and other arthropods conveying diseases or otherwise injurious to man, and
to animals of significance to man. Mainly informative abstracts, with good
monthly and annual author and subject indexes. There is some overlap with
other CAB journals.

943   Review of Plant Pathology (formerly **Review of Applied Mycology**).
1922–. 12 per year. En. Commonwealth Mycological Institute, Ferry
Lane, Kew, Richmond, Surrey TW9 3AF, England.

Another CAB abstracting journal, covering the world literature on plant
pathology, including material on insects acting as disease vectors, etc. Mainly
informative abstracts, well indexed. Some overlap with other CAB journals.

944   Science Citation Index. 1961–. 1 volume per year, issued in parts. En.
Institute for Scientific Information, 3501 Market Street, University City
Science Center, Philadelphia, Pennsylvania 19104, USA.

Bimonthly issues provide an alphabetic index of authors of references cited in
bibliographies or footnotes in other papers or books – i.e. one may use this
index to identify papers in which a particular author has been quoted.
Another section provides for a subject approach, by indexing significant word
pairs from the references so that, for example, one finds a column headed
'insect' followed by a list of other terms (e.g. 'pheromones') used in references including the first. Against each term is a reference back to the basic
listing. A very useful tool, but needs a certain amount of skill to understand
and use properly.

945   Tropical Diseases Bulletin. 1912–. 12 per year. En. Bureau of
Hygiene and Tropical Diseases, Keppel Street, London WC1E 7HT,
England.

Contains abstracts on most aspects of tropical diseases and their control.
Includes a section on medical entomology but other sections are also of
interest, e.g. protozoal diseases and protozoology. Also includes book reviews
and, occasionally, special (review) articles. Each issue has an author index
and a brief subject index, while a much fuller annual index (including a useful
geographic index) is issued quite some time after the close of the volume year.

946   Zoological Record, Insecta (formerly **Record of Zoological Literature**).
1864–. Irregular. En. BioSciences Information Service (BIOSIS),
2100 Arch Street, Philadelphia, Pennsylvania 19103, USA.

The standard reference work, particularly for taxonomic information, providing an index to the publications that have appeared worldwide during the
year, although publication has slipped so badly in recent years that the work's
value has rather diminished. The Insecta section is issued in parts correspond-
ing to various orders of insects, and within each there are five indexes: the
author index; the subject index; the geographical index; the palaeontological

index, and the systematic index. The full bibliographic details of each paper are given only in the author index, to which reference is made in the other indexes.

GUIDES TO THESES AND RESEARCH

Large numbers of postgraduate theses are produced every year. They are by their nature unpublished, although they may be published in some revised form at a later date. However, they often contain valuable information and the need to trace or even borrow copies often arises.

The following reference works can be helpful to guide the researcher to this end.

947   British Reports, Translations and Theses Received by the British Library Lending Division (Including Material from the Republic of Ireland). A monthly list. British Library (address as entry 950).

British report literature and translations produced by British government organizations, industry, universities and learned institutions. The work includes doctoral theses accepted by British universities after 1970. These are available for loan, in most cases under strict copyright conditions. Some are available only in microfiche form.

948   Dissertation Abstracts International, B (Sciences and Engineering). Published 1952–69 as *Dissertation Abstracts.*

The citations are arranged under broad subject headings. Each entry gives full author and title reference and details of the university and degree awarded. A good brief abstract is also included. Prior to 1969 this publication was restricted to doctoral dissertations from the USA but it has now been extended to an international scale. Copies of the theses are available from University Microfilms in microform (usually fiche) or hard copy. A most important reference work for this type of literature.

949   Index to Theses Accepted for Higher Degrees by the Universities of Great Britain and Ireland and the Council for National Academic Awards. Edited by G.H. Paterson and J.E. Hardy. Issued quarterly by Aslib, 3 Belgrave Square, London SW1 8PL.

The date given in the *Index* for the thesis is usually the date of the award of the degree. Aslib do not hold copies of these, but they are usually available on loan from the British Library Lending Division (entry 1046).

950   Research in British Universities Polytechnics and Colleges. Vol. 2, *Biological sciences.* Published annually. British Library Lending Division, Boston Spa, Wetherby, West Yorkshire LS23 7BQ.

This three-volume work is a national register of current research in science and social science. The work is arranged in broad subject headings; within each subject, arrangement is alphabetical by the name or place of the institution. There is a name index and a keyword index.

COMPUTERIZED INFORMATION RETRIEVAL

Anyone involved in entomological research or the planning of pest control programmes needs efficient access to the information contained in the published literature if he or she is to benefit from the results of previous work and avoid duplicating effort.

In the previous section, we described the various abstracting and indexing journals and emphasized their importance as a key to the literature. However, the dramatic increase in the number of potential sources of information, the growing complexity of lines of research, and the sheer volume of literature involved, have combined to make manual searches increasingly time-consuming and ineffective.

A manual search is largely dependent for its success on the quality and depth of the indexing of the abstracts journal being used. Few such journals have indexes that are sufficiently detailed to pinpoint accurately references on the more complex topics, and the searcher normally needs to use the index as a rough guide only, and refer to the abstracts themselves to assess their true relevance. Even if the search is successful, there is no tangible result unless the references found are copied either mechanically or, as is usually the case, longhand onto index cards.

The first step, albeit indirect, towards a solution to these problems came when the major abstracting services began computerizing their journal production processes. Quite apart from enabling the journals to be printed faster and more economically, this change created, as a useful by-product, computer-readable stores or databases of all the information being printed in the journals.

It has been possible for many years to load a database of this kind onto a computer and to program the computer to search for, and print out, records containing specific index terms, keywords or other elements. This type of search, however, has rather limited application and, as there is usually no way of gauging its success before the results are printed, it is in many ways actually inferior to a manual search.

During the 1960s, prompted to some extent by the enormous information explosion in the sciences, particularly in aerospace technology, several organizations were putting considerable effort into developing a more flexible and interactive type of system. As a result, there are now a number of highly sophisticated retrieval systems, each providing access to a wide range of databases, including several relevant to entomology.

Abstracts are stored in such a way as to permit the computer to locate and display those containing specific words, index terms, authors' names, etc., either singly or in combination. In most cases, searching is not restricted to index terms or keywords but can include any or all parts of the record, including the abstract. As another useful feature, most systems are capable of searching for a wordstem so that a single command can be used to retrieve, for example, references containing the terms *parasite, parasites, parasitic, parasitism* and *parasitology*.

In order to access a chosen database, the user connects a relatively simple electronic keyboard device, via the normal telephone network, to the remote 'host' computer, which then responds almost instantaneously to queries and commands typed in by the user. 'Online' searching, as this process is termed, is interactive, with the user gradually refining his search in response to the number and relevance of the abstracts retrieved and displayed at his terminal. When the user is satisfied with the result, he can have all of the retrieved references printed out for further study.

For users with a continuing interest, there is usually an automatic current-awareness option. A search is carried out in the normal way, and each step recorded in the host computer's memory. Subsequently, as each new batch of records is added to the database, usually monthly, the computer will repeat the search on just these new records, and a printout of the selected abstracts is mailed to the user.

Obviously, there would be little appeal in a system that required its users to learn a complicated computer language, and so most of the systems use very similar, very simple commands. These are easily learned within an hour or two, although it takes a little longer to become adept at their use. All sorts of tricks and short cuts can be used to make searches more effective and cheaper.

The most important advantage of online searching is that it enables a search that might have taken hours or even days to carry out manually to be completed in a matter of minutes. Results can be printed out automatically, leaving the user's time free for interpreting and using the information retrieved. The whole retrieval process is relatively effortless, so that online searching is often described as exhaustive but not exhausting!

The costs of online searching comprise basically the following elements: (a) equipment costs, starting from about £1000 for a simple terminal; (b) tele-communications charges, which vary according to the user's geographic location; (c) hourly access charges, which are paid to the system supplier and vary according to the database being used, and (d) print charges, which again are specific to the database.

As a very rough guide, the total cost of searching most of the databases listed in the following pages is at present about £1 per minute for users within Europe, slightly less in the USA, and rather more elsewhere. Although this may sound rather expensive, it should be remembered that a typical search takes only 10 minutes or less, so that the overall cost is relatively low.

Precise details of costs and the procedures to be followed for gaining access can be obtained from the appropriate system supplier, who will also be able to arrange training if required. A list of suppliers' addresses is appended. An alternative to buying or renting the equipment so as to carry out searches directly is to use the services of an intermediary organization. Online searches of any of the databases listed here are available at a moderate charge through the Library/Information Service of the Commonwealth Institute of Entomology (address given under entry 961). Would-be users in university departments may find that their own libraries offer a subsidized service.

Further reading

In addition to the largely excellent user manuals published by the individual systems suppliers and database publishers, the following works are well worth reading.

951   Hall, J.L. 1977. *On-line information retrieval sourcebook.* 267pp. Aslib, London.

A useful overall introduction and reference guide to online searching and the various systems and databases.

952   Hall, J.L. and **Brown, M.J.** 1981. *On-line bibliographic databases: an international directory.* 213pp. Aslib, London.

A useful directory of available databases, although the statistical information given for each will date very rapidly.

953   Harvey, S. 1979. *CAB/CAIN evaluation project: a comparative study on the performance of two agricultural databases in a computerized current awareness service.* 93pp. Centre for Agricultural Publishing and Documentation (PUDOC); Wageningen, Netherlands. (British Library Research and Development Report No. 5483.)

A very thorough and impartial comparison of these two agricultural databases. (CAIN is now known as AGRICOLA.)

Systems suppliers

The following organizations operate the host computers that provide access to the databases listed. They can provide information on the procedures and formalities to be followed in order to gain access to their particular systems, but queries concerning any specific database are best channelled to the appropriate database producer.

Intending users should remember that the geographic location of the host computer is usually of very little consequence to them, as most of the systems listed here can be accessed with varying ease from most parts of the world.

Although several of the suppliers listed do have overseas agents, we have given only the main headquarters address for each, as there is a tendency for agents and their addresses to change quite frequently.

954   BLAISE
2 Sheraton Street, London W1V 4BH, England.

955   DIALOG
DIALOG Information Retrieval Service, Marketing Dept., 3460 Hillview Avenue, Palo Alto, California 94304, USA.

956   ORBIT
SDC Search Service, System Development Corporation, 2500 Colorado Avenue, Santa Monica, California 90406, USA.

957   QUEST
European Space Agency, Space Documentation Service, via Galileo Galilei, I-00044 Frascati, Italy.

Databases of particular relevance to entomology

958   AGRICOLA (*Agric*ultural *O*n-*L*ine *A*ccess)

Printed equivalent: *Bibliography of Agriculture*
Time-span: 1970–
Size: approx. 2 045 000 references at December 1982
Growth rate: approx. 130 000 references per year
Database publisher: National Agricultural Library, USDA SEA/TIS, Beltsville, Maryland 20705, USA
Access systems: DIALOG, ORBIT
Coverage: all aspects of agriculture including applied entomology
Comments: quite a useful general agricultural database, but it does include quite a high proportion of extension literature, which can be rather irritating if one is primarily interested in scientific work. Includes some abstracts, but references are mainly title-only

959   AGRIS (International Information System for Agricultural Sciences and Technology)

Printed equivalent: *Agrindex*
Time-span: 1975–
Size: approx. 620 000 references at December 1982
Growth rate: approx. 120 000 references per year
Database publisher: Food and Agriculture Organization of the United Nations, via delle Terme di Caracalla, I-00100 Roma, Italy
Access systems: QUEST (and by direct telephone dial-up to the International Atomic Energy Agency computer in Vienna)
Coverage: all aspects of agricultural science and technology, including applied entomology
Comments: Another general agricultural database, again with a fairly high proportion of extension literature. Good coverage of report literature, particularly for Asia. Some abstracts, but mainly a title-only database. Rather complicated and difficult to use

960   BIOSIS PREVIEWS (BioSciences Information Service)

Printed equivalent: *Biological Abstracts* and *Biological Abstracts RRM* (formerly *Bioresearch Index*)
Time-span: 1969–
Size: approx. 4 065 000 references at December 1982
Growth rate: 275 000 references per year
Database publisher: Biological Abstracts, 2100 Arch Street, Philadelphia, Pennsylvania 19103, USA
Access systems: DIALOG, ORBIT, QUEST
Coverage: all types of scientific and semi-popular literature on all aspects of life sciences, including pure and applied entomology
Comments: A good database, particularly for searches on taxonomy. The majority of the references are title-only, although abstracts are now being included for many of the more recent additions. Hierarchical indexing aids the retrieval of references by use of broad search terms

961   CAB ABSTRACTS

Printed equivalent: the combined content of all CAB abstracting journals, including *Review of Applied Entomology,* plus *Apicultural Abs-
tracts.*
Time-span: 1973–
Size: approx. 1 452 000 references at December 1982
Growth rate: approx. 150 000 references per year
Database publisher: Commonwealth Agricultural Bureaux, Farnham
House, Farnham Royal, Slough SL2 3BN, England. Queries concerning the entomological content of the database should be addressed to: Librarian/Information Officer, Commonwealth Institute
of Entomology, 56 Queen's Gate, London SW7 5JR, England
Access systems: DIALOG, QUEST
Coverage: all aspects of agricultural science, including related areas of
applied biology
Comments: Quite the best database for applied entomology and apiculture. Easy to use, and the majority of references include an informative abstract

962   CRIS/USDA (Current Research Information System)

Printed equivalent: none
Time-span: 1974–
Size: approx. 47 000 records at December 1982
Growth rate: approx. 24 000 records per year
Database publisher: USDA, Science and Education Administration/
TIS, National Agricultural Library Building, Beltsville, Maryland
20705, USA
Access systems: DIALOG
Coverage: not a bibliographic database, but a collection of factual data
on agriculturally related (many entomological) research projects. The
projects described cover current research in agriculture and related
sciences, sponsored or conducted by USDA research agencies, state
agricultural experiment stations, state forestry schools, and other
cooperating state institutions
Comments: Very useful, particularly in conjunction with a retrospective
search using one of the major bibliographic databases. Many of the
projects included are not written up for publication for quite some
time, if at all, so that the information contained may not be available
elsewhere

963   EXCERPTA MEDICA

Printed equivalent: *Excerpta Medica*
Time-span: 1974–
Size: approx. 1 583 000 references at December 1982
Growth rate: approx. 240 000 references per year
Database publisher: Excerpta Medica, PO Box 1126, 1000 Amster-
dam BC, Netherlands
Access systems: DIALOG

Coverage: human biomedicine and related disciplines, including medical entomology

Comments: Very up-to-date, useful for searches on certain aspects of medical entomology, particularly epidemiology, parasitology and immunology

964 IRL LIFE SCIENCES COLLECTION

Printed equivalent: the combined content of IRL's abstracting journals, including *Entomology Abstracts*

Time-span: 1978–

Size: approx. 455 000 references at December 1982

Growth rate: approx. 130 000 references per year

Database publisher: Information Retrieval Inc., Suite 815, Fisk Building, 250 West 57th Street, New York, NY 10019, USA

Access systems: DIALOG

Coverage: animal behaviour, biochemistry, ecology, entomology (pure and applied), genetics, immunology, microbiology, toxicology and virology, etc.

Comments: Still rather a small database to be of much use for retrospective searches. Good comprehensive coverage of entomology, although the abstracts tend towards the indicative rather than informative

965 MEDLINE (MEDLARS On-Line; MEDLARS = Medical Literature Analysis and Retrieval System)

Printed equivalent: *Index Medicus* and others

Time-span: 1966–

Size: approx. 3 500 000 references at December 1982

Growth rate: approx. 250 000 references per year

Database publisher: US National Library of Medicine, 8600 Rockville Pike, Bethesda, Maryland 20014, USA

Access systems: DIALOG

Coverage: virtually the whole field of biomedicine

Comments: Useful for searches on medical entomology and related topics. A large proportion of the references are now accompanied by abstracts

966 ZOOLOGICAL RECORD

Printed equivalent: *Zoological Record*

Size: approx. 60 000 references at December 1982

Growth rate: approx. 60 000 references per year

Database publisher: BIOSIS, 2100 Arch Street, Philadelphia, Pennsylvania 19103, USA

Access systems: DIALOG

Coverage: virtually all areas of zoology, with emphasis strongly on taxonomic literature

Comments: This database became available online literally as this book went to press, and we have therefore been unable to evaluate it. However, readers may find it useful to refer to item 946 for an indication of the scope and value of the printed version of *Zoological Record*

Example of an online computer search

This example uses the CAB ABSTRACTS database via the Lockheed DIALOG computer system in Palo Alto, California. Search request: Find references on the biological control of the Colorado beetle.

```
A) ENTER YOUR DIALOG PASSWORD
   XXXXXXXX  LOGON FILE50 WED 19AUG81 12:38:27 PORT01D

B) FILE50*:CAB ABSTRACTS - 72-81/JUN
            SET ITEMS DESCRIPTION
            --- ----- -----------

C) ? SELECT (COLORADO(W)BEETLE? OR LEPTINOTARSA(1S)DECEMLINEATA) AND (BIOLOGICAL(W)
   CONTROL OR NATURAL(W)ENEM? OR PARASIT? OR PREDATOR?)
                280  COLORADO(W)BEETLE?
                635  LEPTINOTARSA(1S)DECEMLINEATA
               7450  BIOLOGICAL(W)CONTROL
               6180  NATURAL(W)ENEM?
              54167  PARASIT?
               5256  PREDATOR?
D)          1   140  (COLORADO(W)BEETLE? OR LEPTINOTARSA(1S)DECEMLINEATA) AND (BI
   AL(W)CONTROL OR NATURAL(W)ENEM? OR PARASIT? OR PREDATOR?)
E) ? TYPE 1/7/1

   1/7/1
   1315491    E0069-03329   3
     A   MICROSPORIDIAN  INFECTION  IN  OTIORHYNCHUS  EQUESTRIS  (COLEOPTERA,
   CURCULIONIDAE).
     HOSTOUNSKY, Z. ; WEISER, J.
     ENTOMOLOGICKY USTAV CSAV, FELMINGOVO NAM.2,  160  00  PRAGUE  6-DEJVICE,
   CZECHOSLOVAKIA.
     VESTNIK CESKOSLOVENSKE SPOLECNOSTI ZOOLOGICKE,   1980,   44,3,   160-165
     LANGUAGES: EN
     9 REF., 1 PL., 1 FIG.
     TWO SPECIES OF NOSEMA, WHICH ARE HERE DESCRIBED AS N. ADJUNCTA SP.N.  AND
   N.  EQUESTRIS SP.N., WERE FOUND IN A SINGLE ADULT FEMALE  OF  OTIORHYNCHUS
   EQUESTRIS  (RICHT.)  COLLECTED  FROM  A  HOUSE  IN  EASTERN  BOHEMIA,
   CZECHOSLOVAKIA,  IN THE COURSE OF A SEARCH FOR MICROSPORIDIA THAT MIGHT  BE
   OF  USE  FOR  THE  CONTROL  OF  LEPTINOTARSA  DECEMLINEATA  (SAY).  THE NEW
   MICROSPORIDIA WERE FOUND IN THE  CONNECTIVE  TISSUE,  MALPIGHIAN  TUBULES,
   OENOCYTES  AND FAT-BODY OF OTIORHYNCHUS AND,  WHEN THEY WERE USED TO INFECT
   LARVAE OF L.  DECEMLINEATA AND  GASTROPHYSA  VIRIDULA  (DEG.)  (GASTROIDEA
   VIRIDULA), THEY INFECTED THE SAME TISSUES, CAUSING THE DEATH OF THE LARVAE.

F) ? LOGOFF
              19AUG81 12:47:02 USER6329
   $3.33  0.095 HRS FILE50* 10 DESCRIPTORS
   $0.19  1 TYPES
   $3.52  ESTIMATED TOTAL COST

   LOGOFF 12:47:10
```

See explanatory notes overleaf.

EXPLANATORY NOTES

A) A connection is established with the DIALOG computer by means of the normal telephone network. This 'logging on' procedure takes about 20 seconds on average.

B) All online users have a 'default file', a database to which the computer will normally connect them automatically unless instructed otherwise. In this case, our default file happens to be CAB ABSTRACTS, which is the database we wish to search.

C) DIALOG prompts us with a '?' to indicate that it is ready to accept our commands.

   When searching, it is usually advisable to include possible variant terms, so both scientific and common names are selected for the insect. Similarly, 'biological control' will not necessarily retrieve all relevant references, so the terms 'natural enemies', 'parasites' and 'predators' are also selected

   DIALOG will search for words with a common stem so, for example, we have selected 'parasit?' to retrieve references including the terms *parasite, parasites, parasitic, parasitism* and *parasitology*.

   Online searching uses the principles of Boolean logic, so that search terms can be combined using the logical operators 'and', 'not' and 'or'.

   In this particular example we have used two of these operators, and also some other devices (e.g. the '(W)', which indicates that the words it links must be adjacent), and have also bracketed the two groups of terms together for speed and convenience – rather like an algebraic expression.

D) DIALOG searches the entire database (well over 1 000 000 abstracts) for each term in isolation, then carries out the combination we have requested in order to produce a final total of 140 references in set number 1.

E) We have asked DIALOG to type out just the most recently added of the references, in a format which includes the abstract and full bibliographic citation. We could ask for more of the references to be typed out at our terminal if required, or a simple 'print' command could be used to generate a printout in Palo Alto, for despatch by mail the next day.

F) We have told DIALOG to disconnect us ('logoff'), and this is confirmed together with an account of the time we have spent on the system and the cost (excluding telecommunications). The total time for this particular search was just 0.095 hours (5.7 minutes).

LOCATING SOURCE DOCUMENTS

Although the majority of our readers are probably based at or near a library with at least some entomological content, there will be some who are not, and many whose local resources are strictly limited. Even the largest of libraries is unlikely to be able to meet all the bibliographic needs of an entomological researcher and so, sooner or later, either he or his librarian will need to embark on the tracing of some of the more esoteric literature.

Any key or guide to the literature of a subject is, in our opinion, of minimal use unless it also directs its readers to sources from which that literature may actually be obtained, or at which it may be consulted. In this section, we have attempted to meet this requirement in two ways. First, we have listed a selection of directories and catalogues that may be helpful in identifying sources of specific journals and certain other types of literature. Secondly, we have given a fairly selective international list of libraries that hold significant (either quantitatively or qualitatively) collections of entomological literature. Those who wish actually to purchase new or antiquarian books will find a number of specialist booksellers listed in the section on suppliers (p. 56).

There follows a list of directories, catalogues, and similar sources.

967 *AGLINET Union list of serials.* 1979. 364pp. Food and Agriculture Organization of the United Nations; Rome.

Lists some 6800 titles (then) currently being indexed in *Agrindex*, together with an indication of holdings among the 17 libraries in the international AGLINET network. Includes many titles relevant to economic entomology.

968 *British Union Catalogue of Periodicals Incorporating 'World list of scientific periodicals'. New Periodical Titles.* Butterworths; London.

Issued quarterly, with annual cumulations and other periodic consolidations, *BUCOP* is a valuable continuation of *World list* (*see* entry 974).

969 *Catalogue of the books, manuscripts, maps and drawings in the British Museum (Natural History).* 1903–40. 8 volumes. Trustees of the British Museum (Natural History); London.

Includes the categories of literature stated, plus entries for reprints from publications not represented in the collection. The bulk of the entries are arranged under authors' names, and anonymous works are entered under the principal word or words of the title. Societies and corporate bodies are considered to be the authors of their publications; accounts of surveys and explorations appear under the name of the issuing country; maps under the name of the country charted. A very useful catalogue in view of the Museum's wealth of early entomological literature.

970 *Catalogue of the library of the Royal Entomological Society of London.* 1979. 5 volumes. G.K. Hall; Boston, Massachusetts.

A photographic reproduction of the Society's card index, comprising an author catalogue with title entries for anonymous or edited works. Contains an estimated 78 500 entries for a total of 9000 monographic works, 50 000 pamphlets and 600 journals, some items dating as far back as *c.*1609. There is no subject approach, and there are restrictions on the use of the Society's library, but this is nevertheless an important (if somewhat expensive) catalogue for locating many rare works.

971 *A catalogue of the works of Linnaeus (and publications more immediately relating thereto) preserved in the libraries of the British Museum*

(Bloomsbury) and the British Museum (Natural History) (South Kensington). 1933. 2nd edn. Trustees of the British Museum; London.

A useful catalogue in many ways, and an impressive testament to Linnaeus's work.

972 *List of journals indexed by the National Agricultural Library 1974–76.* 1977. 2 volumes. National Agricultural Library, Beltsville, Maryland.

Covers over six thousand of the serials held by NAL and, as well as giving NAL call numbers (shelf numbers), a good deal of bibliographic information is included for each title. The alphabetic sequence is augmented by geographic and subject indexes. Covers the whole field of agriculture, including many titles relevant to applied entomology.

973 *Serial publications in the British Museum (Natural History) library.* 1980. 3rd edn. 3 volumes. Trustees of the British Museum (Natural History); London.

Lists the titles of serials held in the general and departmental libraries of the British Museum (Natural History) and the Zoological Museum, Tring. Includes expedition reports and other works commonly known by their titles and issued in parts over a period of time, in addition to periodicals in the stricter sense of the term. Entries are arranged alphabetically, directly under the title and mainly in the language of the country of origin. Each title is followed by its standard abbreviation, details of library holdings and locations. An extremely useful list, both for locating material held by the BM(NH) and for checking title abbreviations.

974 *World list of scientific periodicals.* 4th edn. 3 volumes. Butterworths; London.

This edition, published between 1963 and 1965, covers some sixty thousand periodicals which were active between 1900 and 1960. Arrangement is alphabetical, and each entry includes the correct standardized abbreviation, together with details of holdings in the many UK libraries that cooperated. A fairly large number of regularly published conference proceedings are also included, and the whole work is extensively cross-indexed. A particularly useful feature is the signposting to former and successor titles. A later cumulation, covering periodicals active from 1960 to 1968, was published in 1970 as both *World list of scientific periodicals* and *British Union catalogue of periodicals.* These two titles are now combined in the latter (*see* entry 968).

LIBRARIES HOLDING ENTOMOLOGICAL LITERATURE

We have included wherever possible a date of establishment for each of the libraries listed, as a very broad indication of the chronological extent of its collection. It should however be remembered that many libraries hold material published long before they became established.

An approximate indication of size is given for most of the libraries but, in the case of those not limited solely to entomology, these figures relate to the library as a whole unless otherwise stated. All of these figures are of course apt to date very quickly, and are therefore quite emphatically included only as a guide. No attempt has been made to list either the particular services available, or any special restriction on their use, for the simple reason that these have a tendency to change quite frequently.

It is our hope that this list may be useful in suggesting possible sources for some of the more obscure entomological literature. In the first instance, however, readers are urged to try some of the larger, more specialized libraries, such as the Entomology Library of the British Museum (Natural History), the Commonwealth Institute of Entomology, the Hope Department, or the Royal Entomological Society (all in the UK), or the Muséum Nationale d'Histoire Naturelle in France. All of these have particularly fine collections in their respective fields, details of which are included in the appropriate section of the list.

If all the sources listed here fail to provide the solution to a particular bibliographic problem, the reader may need to widen the scope of his search by consulting one of the directories of libraries, of which the following is of particular note:

975 *World guide to libraries.* 1980. 5th edn. K.G. Saur; Munich, New York, London and Paris.

Includes some 37 000 libraries from 157 countries. Arranged geographically, with a broad subject index. A very useful comprehensive guide.

AUSTRALIA

NEW SOUTH WALES

976 **Biological and Chemical Research Institute.** Cnr Victoria Road and Pemberton Street, Rydalmere, New South Wales 2116. (PMB 10, Rydalmere, New South Wales 2116)

Established: 1961

Stock: *c.* 3500 books, 800 periodical titles (450 current)

Main subjects: chemistry, biology and entomology relating to agriculture

977 **Commonwealth Scientific and Industrial Research Organization.** McMaster Laboratory, Parramatta Road, Glebe, New South Wales 2037. (Private Bag No. 1, Glebe, New South Wales 2037)

Established: 1932

Stock: *c.* 2500 books, 650 periodical titles

Main subjects: biological sciences, bacteriology, entomology, microbiology, parasitology, biochemistry, zoology, veterinary physiology, veterinary science, agriculture, animal husbandry. Houses the library of the Australian Veterinary Association

978 **University of Sydney.** School of Public Health and Tropical Medicine Library, University of Sydney, New South Wales 2006.

Established: 1910

Stock: *c.* 28 000 books, 1100 periodical titles, 1500 WHO monographs, reports and public health papers

Main subjects: public health and tropical medicine, preventive and social medicine, nutrition, bacteriology, biochemistry, radiology, occupational and environmental health, parasitology and entomology

NORTHERN TERRITORY

979   Department of the Northern Territory. Animal Industry and Agriculture Branch, Block 1, Mitchell Street, Darwin, Northern Territory 5790. (PO Box 5150, Darwin, Northern Territory 5790)

Established: 1945
Stock: *c*. 1500 books, 600 periodical titles, plus pamphlets and trade catalogues
Main subjects: agriculture, animal production, veterinary science, botany, entomology, chemistry, land conservation

QUEENSLAND

980   Bureau of Sugar Experiment Stations. 99 Gregory Terrace, Brisbane, Queensland 4000. (PO Box 292, North Brisbane, Queensland 4000)

Established: 1930
Stock: *c*. 2500 books, 200 periodical titles, plus sugar conference and congress reports
Main subjects: sugar cane agriculture, sugar technology, analytical chemistry, soils and agronomy, entomology, plant physiology, plant pathology, farm management and extension

981   Department of Primary Industries. Darling Downs Library, Tor Street, Toowoomba, Queensland 4350. (PO Box 102, Toowoomba, Queensland 4350)

Established: –
Stock: *c*. 1000 books, 170 periodical titles
Main subjects: agriculture, soil conservation, entomology, plant pathology, dairying, chemistry

982   Queensland Institute of Medical Research. Library, Bramston Terrace, Herston, Queensland 4006.

Established: 1947
Stock: *c*. 14 000 volumes, 190 current periodical titles
Main subjects: medical research, including medical entomology

SOUTH AUSTRALIA

983   Department of Agriculture and Fisheries. Agriculture Library, 25 Grenfell Street, Adelaide, South Australia 5000. (GPO Box 1671, Adelaide, South Australia 5001)

Established: 1920s
Stock: *c*. 4600 books, 500 periodical titles
Main subjects: veterinary science, agriculture and agronomy, entomology, economics, statistics, history, nutrition, chemicals

984   Waite Agricultural Research Institute. Waite Road, Netherby, South

Australia. (All mail to: Private Bag, Post Office, Glen Osmond, South Australia 5064)

Established: 1925

Stock: *c*. 10 000 books, 3660 periodical titles

Main subjects: general agriculture, animal sciences, biometry, entomology, agricultural biochemistry, general biochemistry, soil science, plant physiology, plant pathology, agronomy and plant breeding

TASMANIA

985 Department of Agriculture, Tasmania. New Town Research Laboratories, St Johns Avenue, New Town, Tasmania 7008.

Established: –

Stock: *c*. 2500 books, 1550 periodical titles (figures include branch library holdings)

Main subjects: agriculture and related fields, primarily entomology, plant pathology, horticulture, agronomy, biometrics

986 Royal Society of Tasmania. 2nd Floor, Morris Miller Library, University of Tasmania, Tasmania 7000. (GPO Box 1168 M, Hobart, Tasmania 7001)

Established: 1845

Stock: *c*. 4250 books, 1825 periodical titles

Main subjects: agriculture, biology, chemistry, botany, entomology, geology, medicine, mineralogy, physics, palaeontology, zoology

VICTORIA

987 Department of Agriculture, Victoria. Plant Sciences Library, Victorian Plant Research Institute, Swan Street, Burnley, Victoria 3121.

Established: 1895

Stock: *c*. 12 000 books, 300 periodical titles, 35 000 pamphlets

Main subjects: plant pathology, entomology, nematology, plant virology, horticulture

WESTERN AUSTRALIA

988 Department of Agriculture, Western Australia. Jarrah Road, South Perth, Western Australia 6151.

Established: 1894

Stock: *c*. 10 000 books, 1500 periodical titles, 500 report series, 15 000 pamphlets, 3000 slides

Main subjects: agriculture, botany, entomology, veterinary medicine

BELGIUM

989 Société Royale d'Entomologie de Belgique. IRSN, Bibliothèque, rue Vautier 31, B-1040 Bruxelles.

Established: 1855

Stock: *c*. 16 000 volumes

Main subjects: entomology

CANADA

990 Agriculture Canada, Entomology Research Library. K.W. Neatby Building, Rm 4061, Central Experimental Farm, Ottawa, Ontario K1A 0C6

Established: 1919

Stock: *c*. 11 000 books, 12 000 periodical volumes, 500 current periodical titles

Main subjects: entomology – taxonomy and biology, biosystematics, faunistics, evolution, palaeontology, zoology, ecology, genetics, geology, physiology. Collection includes a substantial number of USDA publications

991 Agriculture Canada, Experimental Farm Library. CP 400, La Pocatière, Québec G0R 1Z0

Established: –

Stock: *c*. 950 books, 1200 periodical volumes, 250 maps, 1200 aerial photographs

Main subjects: plant pathology, mycology, biochemistry, plant physiology, general agriculture, plant breeding, botany, entomology, bacteriology

992 Agriculture Canada, Lethbridge Research Station Library. Lethbridge, Alberta T1J 4B1

Established: 1949

Stock: *c*. 18 000 volumes, 870 current periodical titles

Main subjects: includes entomology and acarology

993 Canada Department of Agriculture, Kentville Research Station Library. Kentville, Nova Scotia B0W 1Y0

Established: 1950

Stock: *c*. 3500 volumes, 175 current periodical titles

Main subjects: includes entomology

994 Canada Department of Agriculture, Vancouver Research Station Library. 6660 NW Marine Drive, Vancouver, British Columbia V6T 1X2

Established: 1959

Stock: *c*. 7000 volumes, 180 current periodical titles

Main subjects: includes entomology

995 Canada Fisheries & Environment, Centre de Recherches Forestières des Laurentides, Library. PO Box 3800, Ste Foy, Québec G1V 4C7

Established: 1952

Stock: *c*. 7000 volumes

Main subjects: includes entomology

996 Canadian Forestry Service, Northern Forest Research Centre, Library. 5320 122nd Street, Edmonton, Alberta T6H 3S5

Established: 1948

Stock: *c*. 6000 volumes, 200 current periodical titles

Main subjects: includes entomology

997 McGill University, Macdonald College Library. Barton Building, Macdonald College Post Office, Ste Anne de Bellevue, Québec, H0A 1C0

Established: 1907

Stock: *c.* 59 000 books, 28 000 periodical volumes, 124 000 government documents

Main subjects: environmental, food and agricultural sciences. Includes the Lyman Entomological Collection (old and rare books)

998 Nova Scotia Museum Library. 1747 Summer Street, Halifax, Nova Scotia B3H 3A6

Established: 1885

Stock: *c.* 10 000 volumes, 140 current periodical titles

Main subjects: zoology, botany, social history of Nova Scotia, marine history. Includes special entomological collection (1000 volumes)

999 Pacific Forest Research Centre, Canada Department of the Environment, Library. 506 West Burnside Road, Victoria, British Columbia V8Z 1M5

Established: 1960

Stock: *c.* 9000 volumes, 300 current periodical titles

Main subjects: includes forest entomology

1000 Ministère de l'Agriculture, Québec Province, Bibliothèque. 200-A Chemin Ste-Foy, Québec, Québec Province G1X 1R4

Established: 1942

Stock: *c.* 10 000 books, 1200 periodical volumes, 750 current periodical titles

Main subjects: agriculture, entomology, veterinary medicine

1001 University of British Columbia, Spencer Entomological Museum, Library. Department of Zoology, Vancouver, British Columbia V6T 1W5

Established: 1953

Stock: *c.* 300 books, 8000 reprints, 50 periodical titles.

1002 University of Manitoba, Agriculture Library. Winnipeg, Manitoba R3T 2N2

Established: 1906

Stock: *c.* 9500 volumes, 290 current periodical titles

Main subjects: agriculture, biology, entomology, agricultural economics, soil science, food science

CZECHOSLOVAKIA

1003 Československá Akademie Věd, Entomologicky Ústav, Knihovna [Czechoslovak Academy of Sciences, Institute of Entomology, Library]. Manesova 55, 120 00 Praha 2.

Established: 1954

Stock: *c.* 5300 volumes

Main subjects: entomology

1004   Československy Svaz Vcelaru, Knihovna [Czechoslovak Union of Beekeepers, Library]. Nove Mesto, Kremencova 8, 110 00 Praha.

Established: 1945
Stock: *c*. 12 600 volumes
Main subjects: apiculture

1005   Slovenská Akadémie Vied, Ustav Experimentalnej Fytopatologie i Entomologie Knižnica [Slovak Academy of Sciences, Institute of Experimental Plant Pathology and Entomology, Library]. Ivanka pri Dunaji, Slovakia.

Established: –
Stock: *c*. 7000 volumes
Main subjects: plant pathology and agricultural entomology

FINLAND

1006   Maatalouden Tutkimuskeskus, Tuhoeläintutkimuslaitos, Kirjasto [Department of Plague Investigation, Agricultural Research Centre]. PL 18, SF-01301 Vantaa 30.

Established: 1898
Stock: *c*. 5000 volumes, 150 current periodical titles
Main subjects: includes relevant aspects of medical and agricultural entomology

1007   Tieteellisten Seurain Kirjasto [Library of Scientific Societies]. Snell-maninkatu 9–11, SF-00170 Helsinki 17.

Established: 1899
Stock: *c*. 7000 volumes, 400 periodical titles
Main subjects: includes special entomological collection

FRANCE

1008   Ecole Nationale Supérieure Agronomique et Centre de Recherches Agronomiques du Midi, Bibliothèque. 9 Place Viala, F-34060 Montpellier

Established: 1872
Stock: *c*. 35 000 volumes, 650 periodical titles
Main subjects: includes entomological special collection

1009   Muséum National d'Histoire Naturelle, Laboratoire d'Entomologie Générale et Appliquée, Bibliothèque. 45 bis rue de Buffon, F-75005 Paris

Established: 1829
Stock: *c*. 30 000 volumes, 150 current periodical titles
Main subjects: pure and applied entomology

1010   Société Entomologique de France, Bibliothèque. 45 rue de Buffon, F-75005 Paris

Established: 1832

Stock: *c*. 10 000 volumes, 130 current periodical titles
Main subjects: entomology

1011   Société Central d'Apiculture, Bibliothèque. 28 rue Serpente, F-75006 Paris

Established: 1856
Stock: *c*. 3000 volumes, 30 current periodical titles
Main subjects: apiculture

GERMAN DEMOCRATIC REPUBLIC

1012   [Entry deleted]

1013   Institut für Pflanzenschutzforschung, Abt. Taxonomie der Insekten, Akademie der Landwirtschaftswissenschaften der DDR, Entomologische Bibliothek. Schicklerstr. 5, DDR-13 Eberswalde

Established: 1886
Stock: *c*. 55 000 volumes, 862 current periodical titles
Main subjects: entomology, taxonomy

FEDERAL REPUBLIC OF GERMANY

1014   Deutsche Entomologische Gesellschaft, Bibliothek. Invalidenstr. 43, Berlin N4 *and also* Konigin Luise Str. 19, Berlin

1015   Münchner Entomologische Gesellschaft, Bibliothek. Maria-Ward-Str. 1b, Schloss Nymphenburg, Nordflugel, D-8000 München 19

Established: 1905
Stock: *c*. 12 000 volumes, 220 current periodical titles
Main subjects: entomology

ITALY

1016   Food and Agriculture Organization of the United Nations. David Lubin Memorial Library, via delle Terme di Caracalla, I-00100 Roma

Established: 1909
Stock: *c*. 1 000 000 volumes, 6000 current periodical titles
Main subjects: all aspects of agriculture, including pest control

1017   Istituto Nazionale di Entomologia, Biblioteca. via Catone 34, I-00192 Roma

Established: –
Stock: *c*. 4000 volumes, 345 current periodical titles
Main subjects: entomology

1018   Stazione di Entomologia Agraria, Biblioteca. via Romana 15–17, I-50125 Firenze

Established: 1875

Stock: *c*. 31 000 volumes, 514 current periodical titles
Main subjects: agricultural entomology

1019 Università di Bologna, Istituto di Entomologia, Biblioteca. via Filippo Re 6, I-40126 Bologna

Established: 1930
Stock: *c*. 13 000 volumes, 800 current periodical titles
Main subjects: entomology

1020 Università di Catania, Istituto di Entomologia Agraria, Biblioteca. via Valdisavoia 5, I-95123 Catania

Established: 1952
Stock: *c*. 3000 volumes, 60 current periodical titles
Main subjects: agricultural entomology

1021 Università degli Studi di Milano, Istituto di Entomologia Agraria, Biblioteca. via Giovanni Celoria 2, I-20133 Milano

Established: –
Stock: *c*. 7000 volumes, 240 current periodical titles
Main subjects: agricultural entomology

1022 Università di Napoli, Facoltà di Agraria, Biblioteca. via dell'Università, Portici, I-80138 Napoli

Established: 1872
Stock: –
Main subjects: includes special entomological collection

1023 Università di Napoli, Istituto di Entomologia e Zoologia Agraria Filippo Silvestri, Biblioteca. via Università 100, I-80055 Portici

Established: 1900
Stock: *c*. 32 000 volumes, 1500 current periodical titles
Main subjects: agricultural entomology, agricultural zoology

1024 Università degli Studi di Palermo, Istituto di Entomologia Agraria, Biblioteca. viale delle Scienze, Parco d'Orléans, I-90128 Palermo

Established: 1943
Stock: *c*. 4000 volumes, 70 current periodicals
Main subjects: agricultural entomology

1025 Università degli Studi di Pisa, Istituto di Entomologia Agraria, Biblioteca. Facoltà di Agraria, via S. Michele degli Scalzi 2, I-56100 Pisa

Established: 1940
Stock: *c*. 7000 volumes, 70 current periodical titles
Main subjects: agricultural entomology

1026 Università degli Studi di Salerno, Istituto di Entomologia Agraria, Biblioteca. via Enrico de Nicola, I-07100 Sassari

Established: 1948
Stock: *c*. 3000 volumes, 350 current periodical titles
Main subjects: agricultural entomology

NETHERLANDS

1027 Nederlandse Entomologische Vereniging, Bibliotheek. Plantage Middenlaan 64, NL-1004 Amsterdam

Established: 1860
Stock: *c.* 18 000 volumes, 700 current periodical titles, 100 000 reprints
Main subjects: entomology, including many old works on systematics

NEW ZEALAND

1028 Entomology Division and Plant Diseases Division, DSIR, Mount Albert Research Centre Library. 120 Mount Albert Road, Private Bag, Auckland

Established: 1920
Stock: *c.* 38 000 volumes, 460 microfilms, 1900 current periodical titles
Main subjects: entomology, plant pathology

NORWAY

1029 Stavanger Museum, Bibliotek. Musegate 16, N-4000 Stavanger

Established: 1877
Stock: *c.* 26 000 volumes, 840 current periodical titles
Main subjects: includes entomology

1030 Universitetet i Oslo, Etnografiske Museum, Biblioteket. Frederiksgate 2, Oslo 1

Established: 1857
Stock: *c.* 25 000 volumes, 300 current periodical titles
Main subjects: includes entomology

PERU

1031 Sociedad Entomológica del Perú, Biblioteca. Apdo 4796, Lima

Established: 1956
Stock: *c.* 2500 volumes
Main subjects: entomology

POLAND

1032 Polskie Towarzystwo Entomologiczne, Biblioteka [Polish Entomological Society Library]. ul. Nowy Swiat 72, 00330 Warszawa

Established: 1920
Stock: *c.* 6000 volumes, 275 current periodicals
Main subjects: entomology

1033 Uniwersytet Imienia Marii Curie-Skłodowskiej, Biblioteka. ul. Narutowicza 4, Lublin

Established: 1944
Stock: –
Main subjects: includes Konstanty Strawinski entomological collection

SOUTH AFRICA

1034 **South African Museum, Library.** Queen Victoria Street, PO Box 61, Cape Town 8000

Established: 1855
Stock: *c*. 14 500 volumes, 28 000 pamphlets, 600 maps, 1060 current periodical titles
Main subjects: includes entomology

SPAIN

1035 **Escuela Tecnica Superior de Ingenieros de Montes, Biblioteca.** Ciudad Universitaria, Madrid 3

Established: 1848
Stock: *c*. 32 500 volumes, 480 current periodicals, 2500 maps
Main subjects: includes entomology

1036 **Instituto Español de Entomología, Biblioteca.** Paseo de la Castellana 84, Madrid
Established: 1940
Stock: *c*. 3000 volumes, 350 current periodical titles
Main subjects: entomology

SWEDEN

1037 **Entomologiska Foreningens Bibliotek.** Riksmuseet, S-101 00 Stockholm

Established: –
Stock: *c*. 15 000 volumes, 280 current periodical titles
Main subjects: entomology

1038 **Naturhistoriska Riksmuseet; Sekt. F, Entomologi, Biblioteket.** Roslagsvagen 120, S-104 05 Stockholm
Established: 1841
Stock: *c*. 16 000 volumes, 200 current periodical titles
Main subjects: entomology

1039 **Universitetsbibliotek Lund.** Helgonabacken, S-221 03 Lund, Sweden
Established: 1666
Stock: –
Main subjects: includes special entomological collection

SWITZERLAND

1040 Eidgenössische Technische Hochschule Zürich, Entomologisches Institut, Bibliothek. Clausiusstr. 21, CH-8006 Zürich
Established: 1933
Stock: *c.* 2300 volumes, 70 current periodical titles
Main subjects: entomology

1041 Eidgenössische Versuchsanstalt für Obst-, Wein- und Gartenbau. Station Fédérale de Recherches en Arboriculture, Viticulture et Horticulture, Bibliothek, Schloss, CH-8820 Wadenswil
Established: 1890
Stock: *c.* 8200 volumes, 500 current periodical titles, 8000 slides, theses, patents
Main subjects: includes agricultural/horticultural entomology

1042 Schweizerische Entomologische Gesellschaft, Bibliothek. Leonhardstr. 33, CH-8006 Zürich
Established: 1858
Stock: *c.* 18 100 volumes
Main subjects: entomology

1043 Verein Deutschschweizer Bienenfreunde, Bibliothek. Rosenberg, CH-6300 Zug
Established: 1928
Stock: *c.* 2500 volumes
Main subjects: includes entomology

TRINIDAD AND TOBAGO

1044 Commonwealth Institute of Biological Control, Library. Gordon Street, Curepe, Trinidad
Established: 1962
Stock: *c.* 5500 volumes
Main subjects: biological control and related aspects of entomology and biology

UK

1045 Animal Virus Research Institute. Pirbright, Woking, Surrey, England
Established: *c.* 1950
Stock: *c.* 1400 books, 100 current periodical titles
Main subjects: virology and allied fields, including veterinary science, biochemistry, genetics and entomology

1046 British Library Lending Division. Boston Spa, Wetherby, West Yorkshire LS23 7BQ, England
Established: 1973

Stocks: *c*. 3 000 000 volumes, 2 000 000 microforms, 270 000 theses, 41 000 translations, 50 000 current periodical titles
Main subjects: all subjects, including entomology and related fields. Collection includes all US dissertations appearing in *Dissertation Abstracts A* and *B* from 1970 onwards, and doctoral theses from most British universities

1047 British Medical Association, Nuffield Library. BMA House, Tavistock Square, London WC1H 9JP, England
Established: 1887
Stock: *c*. 85 000 volumes, 1000 current periodical titles
Main subjects: all fields of medicine, including medical entomology

1048 British Museum (Natural History), Entomology Library. Cromwell Road, London SW7 5BD, England
Established: 1881
Stock: *c*. 76 000 volumes, 650 current periodical titles
Main subjects: systematic and economic entomology. Collection includes many rare works, manuscripts, drawings and scientific collectors' notebooks.

1049 Central Veterinary Laboratory, Library. Ministry of Agriculture, Fisheries and Food, New Haw, Weybridge, Surrey KT15 3NB, England
Established: 1917
Stock: *c*. 65 000 items, 980 curent periodical titles
Main subjects: all aspects of veterinary science, including veterinary entomology

1050 Centre for Overseas Pest Research. College House, Wrights Lane, London W8 5SJ, England
Established: 1971 (1928–70 as Antilocust Library)
Stock: *c*. 75 000 books, memoirs, reprints, pamphlets and theses, 10 000 photographs, 4000 slides, 1300 current periodical titles
Main subjects: research on overseas pest control and related subjects, with special emphasis on Orthoptera (grasshoppers and locusts)

1051 Commonwealth Institute of Entomology, Library. 56 Queen's Gate, London SW7 5JR, England
Established: 1913
Stock: *c*. 35 000 volumes, 70 000 reprints, 900 current periodical titles
Main subjects: applied entomology, including all insects, mites and related arthropods of economic importance

1052 Hope Department of Entomology, University Museum, Library. Oxford OX1 3PW, England
Established: 1849

Stock: *c*. 11 500 volumes, 52 000 reprints, 2000 slides, 150 current
   periodical titles
Main subjects: entomology. Includes many rare works

1053 **H.K. Lewis & Co. Ltd, Medical, Scientific and Technical Library.** 136
Gower Street, London WC1E 6BS, England

Established: –
Stock: –
Main subjects: medicine, surgery, and pure and applied science,
   including entomology and related fields. Collection includes only the
   latest edition of works within its scope, and is particularly good for
   American works. This is a subscription library, run by a bookseller

1054 [Entry deleted]

1055 **Institute of Terrestrial Ecology, Monks Wood, Library.** Natural Environment Research Council, Monks Wood Experimental Station,
Abbots Ripton, Huntingdon PE17 2LS, England

Established: 1963
Stock: *c*. 3600 books, 270 current periodical titles
Main subjects: biological sciences relevant to terrestrial ecological systems, and environmental protection and management, including
   material on insects and other invertebrates

1056 **Liverpool School of Tropical Medicine, Library.** Pembroke Place,
Liverpool L3 5QA, England

Established: 1898
Stock: *c*. 14 500 volumes, 20 000 reprints, 260 current periodical titles
Main subjects: tropical medicine, including medical entomology and
   related fields

1057 **London School of Hygiene and Tropical Medicine, Library.** University
of London, Keppel Street, London WC1E 7HT, England

Established: 1924
Stock: *c*. 60 000 volumes, 1200 current periodical titles
Main subjects: preventive medicine and microbiology (temperate and
   tropical). Includes a considerable amount of literature on medical
   entomology and related fields. Specializes in epidemiological
   reports.

1058 **Ministry of Agriculture, Fisheries and Food, Slough Laboratory, Library** (formerly Pest Infestation Control Laboratory). London Road,
Slough, Buckinghamshire SL3 7HJ, England

Established: –
Stock: *c*. 7300 volumes, 20 000 pamphlets, 350 current periodical
   titles
Main subjects: stored-product and household pests (insects and mites),
   pesticides (including physiological means of control), stored products, microbial associations with stored products, and with stored-
   products pests

1059 Ministry of Agriculture, Fisheries and Food, Harpenden Laboratory, Library (formerly Plant Pathology Laboratory). Hatching Green, Harpenden, Hertfordshire AL5 2BD, England

Established: 1918

Stock: *c*. 5000 books, 3500 pamphlets, 3000 photographs (slides and prints), 550 current periodical titles

Main subjects: plant health, pests and diseases of plants, agricultural pesticides

1060 Medical Research Council, Library. National Institute for Medical Research, The Ridgeway, Mill Hill, London NW7 1AA, England

Established: 1949

Stock: *c*. 34 000 volumes, 25 000 pamphlets, 670 current periodical titles

Main subjects: biology, chemistry, biochemistry, biophysics, genetics, physiology, pharmacology, medicine, protozoology, parasitology, microbiology, immunology, endocrinology, veterinary science. Includes medical/veterinary entomology

1061 Nature Conservancy Council, Library. 19/20 Belgrave Square, London SW1X 8PY, England (and at Banbury, Oxfordshire)

Established: 1949

Stock: *c*. 30 000 books and pamphlets, 1500 periodical titles

Main subjects: biological sciences, nature conservation, land use and allied subjects. Includes material on insects and other invertebrates

1062 Rothamsted Experimental Station, Library. Rothamsted Experimental Station, Harpenden, Hertfordshire AL5 2JQ, England

Established: 1843

Stock: *c*. 80 000 volumes, 2000 current periodical titles

Main subjects: agriculture and related topics, including botany, plant pathology, soil science, pedology, nematology, entomology, apiculture, fungicides. The collection is very rich in old and rare works

1063 Royal Entomological Society of London, Library. 41 Queen's Gate, London SW7 5HU, England

Established: 1833

Stock: *c*. 9000 volumes, 50 000 reprints, 250 current periodical titles

Main subjects: entomology. The collection includes many rare works

1064 Royal College of Veterinary Surgeons, Wellcome Library. 32 Belgrave Square, London SW1X 8QP, England

Established: 1844

Stock: *c*. 25 000 volumes, 245 current periodical titles

Main subjects: veterinary science, including veterinary entomology

1065 Tate & Lyle Ltd, Philip Lyle Memorial Research Laboratory, Library. PO Box 68, Reading, Berkshire RG6 2BX, England

Established: 1971

Stock: *c*. 3000 books, 4000 periodical titles

Main subjects: all aspects of sugar refining and technology, history, agriculture, etc., organic chemistry, food and nutrition, industrial chemistry, chemical engineering. Includes one of the most complete collections of sugar conference proceedings, of great importance to applied entomology

1066   Tropical Products Institute, Library. 56–62 Gray's Inn Road, London WC1X 8LU, England

Established: 1894

Stock: *c*. 120 000 items, 1200 current periodical titles, 1000 annual report series

Main subjects: tropical agriculture, forestry and animal products, with particular emphasis on post-harvest aspects. Includes quite a high proportion of economic entomology

USA

CALIFORNIA

1067   California Academy of Sciences, J.W. Mailliard Jr Library. Golden Gate Park, San Francisco, CA 94118

Established: 1853

Stock: *c*. 85 000 volumes (mainly periodicals), 2110 current periodical titles

Main subjects: botany, entomology, invertebrate zoology, geography, geology, herpetology, ichthyology, mammallology, ornithology, marine biology, palaeontology, conservation/ecology

1068   Stauffner Chemical Company, Mountain View Research Center Library. Box 760, Mountain View, CA 94042, USA

Established: –

Stock: *c*. 7000 volumes

Main subjects: includes pest control

1069   Stored-Product Insects Research Laboratory, USDA, SEA. 5578 Air Terminal Drive, Fresno, CA 93727

Established: –

Stock: *c*. 250 books, 360 periodical titles (24 current)

Main subjects: stored-product entomology, insect pathology

1070   University of California, Berkeley, Entomology Library. Wellman Hall, Room 210, Berkeley, CA 94720

Established: 1943

Stock: *c*. 12 300 volumes, 17 000 pamphlets, 340 current periodical titles

Main subjects: entomology, parasitology, helminthology, biological control, invertebrate pathology

1071   University of California, Davis, General Library. Davis, CA 95616

Established: –
Stock: *c*. 14 000 volumes (entomology/nematology)
Main subjects: agriculture, biological sciences, engineering and technology, humanities and social sciences, physical sciences. The entomology collection is a small part of a very large library (800 000 books, etc.)

CONNECTICUT

1072   Connecticut Agricultural Experiment Station, Osborne, Library. 123 Huntington Street, New Haven, CT 06504

Established: 1875
Stock: *c*. 23 000 volumes, 400 current periodical titles
Main subjects: botany, chemistry, entomology

WASHINGTON, DC

1073   Smithsonian Institution, National Museum of Natural History, Entomology Branch Library. Natural History Building, Washington, DC 20506

Established: –
Stock: *c*. 17 500 volumes, 280 current periodical titles
Main subjects: taxonomic and medical entomology. Includes the Casey Collection (Coleoptera)

FLORIDA

1074   Florida Department of Agriculture and Consumer Services, Division of Plant Industry, Library. Box 1269, Gainesville, FL 32601

Established: 1915
Stock: *c*. 9600 volumes
Main subjects: includes entomology (including rare works)

1075   Florida Department of Health and Rehabilitative Services, Florida Medical Entomology Library. Box 520, Vero Beach, FL 32960

Established: –
Stock: *c*. 7000 volumes
Main subjects: medical entomology

1076   Horticultural Research Laboratory Library, USDA SEA. 2210 Camden Road, Orlando, FL 32803

Established: 1970
Stock: *c*. 2500 volumes, 100 current periodical titles, reprints, slides, photographs
Main subjects: citrus culture, citrus breeding, citrus processing, citrus insects, nematology, plant pathology, plant physiology, biochemistry

1077   University of Florida, Hume Library. Institute of Food and Agricultural Sciences, McCarty Hall, Gainesville, FL 32611

Established: 1900
Stock: *c.* 89 500 volumes, 187 000 separates, 4650 microfiche (botany classics)
Main subjects: agriculture, botany, horticulture, entomology, bacteriology, animal science, statistics, biological science

GEORGIA
1078 Stored-Product Insects Research & Development Laboratory, USDA SEA, Library. 3401 Edwin Street, Box 22909, Savannah, GA 31403

Established: –
Stock: *c.* 3200 volumes, 11 500 reprints, 150 current periodical titles
Main subjects: stored-product insect control, entomology, chemistry, biology, insect-resistant packaging, mothproofing, insect rearing

HAWAII
1079 Hawaiian Entomological Association, Experiment Station Library, HSPA. 1527 Keeaumoku Street, Honolulu, HI 96822

Established: 1898
Stock: –
Main subjects: entomology. Includes the Exchange Collection of the Hawaiian Entomological Association

ILLINOIS
1080 University of Illinois, Natural History Survey, Illinois State, Library. 196 Natural Resources Building, Urbana, IL 61801

Established: 1858
Stock: *c.* 33 000 volumes
Main subjects: includes a very substantial proportion of entomology relating principally to Illinois

1081 University of Notre Dame, Life Sciences Research Library. Galvin Life Sciences Building, Notre Dame, IL 46556

Established: 1938
Stock: *c.* 43 000 volumes, 2100 microforms, 650 current periodical titles
Main subjects: biology, genetics, microbiology, entomology, parasitology

MARYLAND
1082 Insect Control and Research Inc., Library. 1330 Dillon Heights Avenue, Baltimore, MD 21228

Established: 1946
Stock: *c.* 1100 volumes, 5000 reprints, 50 current periodical titles
Main subjects: entomology, plant pathology, tropical diseases, pesticides, agriculture, food plant sanitation

1083 National Agricultural Library. US Department of Agriculture, 10301 Baltimore Road, Beltsville, Maryland 20705

Established: 1862
Stock: *c*. 1 600 000 volumes, 4000 current periodical titles
Main subjects: historical agriculture, poultry, etc. Includes important literature on economic entomology

MICHIGAN

1084 Michigan State University, Special Collections Division. University Library, East Lansing, MI 48824

Established: 1960
Stock: 100 000 volumes, 150 current periodical titles
Main subjects: includes veterinary medicine (history) – 1523 items, entomology

MINNESOTA

1085 University of Minnesota, St Paul, Entomology Library. 1980 Folwell Avenue, 375 Hodson Hall, St Paul, MN 55108

Established: 1905
Stock: *c*. 26 000 volumes, 28 000 separates, 198 cassettes, 790 current periodical titles
Main subjects: entomology, fisheries, wildlife, pesticides, water pollution, limnology. Includes special Bee Collection (800 monographs, 87 journal titles)

MONTANA

1086 US National Institutes of Health, Rocky Mountain Laboratory Library. Pine Grove Addition, Hamilton, MT 59840

Established: 1929
Stock: *c*. 36 000 volumes
Main subjects: entomology, particularly medical aspects

NEBRASKA

1087 University of Nebraska, C.Y. Thompson Library. East Campus, Lincoln, NE 68583

Established: –
Stock: *c*. 172 000 volumes, 1375 microforms, 4942 current periodical titles
Main subjects: agriculture, home economics, textiles, wildlife conservation, human development, nutrition, applied sciences. Includes special Entomology Collection containing many rare volumes

NEW JERSEY

1088 American Cyanamid Company, Agricultural Division, Technical Information Center. Box 400, Princeton, NJ 08540

Established: 1947
Stock: *c*. 28 000 volumes, 200 dissertations, pamphlets, patents, 650 current periodical titles

Main subjects: agronomy, entomology, veterinary medicine, agricultural chemicals. Collection includes significant holdings of USDA and Agricultural Experiment Station publications

1089 S.B. Penick & Co., Division of CPC International, Research Library. 215–225 Watchung Avenue, Orange, NJ 07050

Established: 1920
Stock: *c.* 5500 volumes
Main subjects: includes agricultural entomology

NEW YORK

1090 American Museum of Natural History, Library. Central Park W. at 79th Street, New York, NY 10024

Established: 1903
Stock: 225 000 volumes, films, maps, photographs, manuscripts, memorabilia, rare books, 6000 current periodical titles
Main subjects: anthropology, archaeology, palaeontology, entomology, mammalology, ornithology, ichthyology, malacology, herpetology, mineralogy, geology, museology, travels and voyages

1091 Cornell University, Comstock Memorial Library of Entomology. Comstock Hall, Ithaca, NY 14853

Established: 1914
Stock: *c.* 27 000 volumes
Main subjects: entomology

NORTH CAROLINA

1092 North Carolina State University, D.H. Hill Library. Box 5007, Raleigh, NC 27607

Established: 1899
Stock: *c.* 741 000 volumes, 26 800 maps and atlases, 1 400 000 microforms, 539 000 US government publications, 42 000 slides, 12 900 current periodical titles
Main subjects: engineering, agriculture, forestry, textiles, architecture, biological sciences, genetics, statistics. Includes the Tippmann Collection (entomology)

1093 US Navy, Naval Medical Field Research Laboratory, Scientific Library. Camp Lejeune, NC 28542

Established: 1943
Stock: *c.* 16 000 volumes
Main subjects: includes medical entomology

OHIO

1094 Lloyd Library and Museum. 917 Plum Street, Cincinnati, OH 45202

Established: –

Stock: *c*. 165 000 volumes, 120 000 pamphlets, 1500 current periodical titles
Main subjects: botany, pharmacy, biology, chemistry, natural science,
zoology, entomology, mycology, eclectic medicine

PENNSYLVANIA

1095 Academy of Natural Sciences of Philadelphia, Library. 19th Street &
Parkway, Philadelphia, PA 19103

Established: 1812
Stock: *c*. 175 000 volumes
Main subjects: natural sciences, including entomology

**1096 Pennsylvania State University, Frost Entomological Museum, Tax-
onomic Research Library.** 106 Patterson Building, University Park,
PA 16802

Established: 1972
Stock: –
Main subjects: taxonomic entomology

WEST VIRGINIA

**1097 West Virginia State Department of Agriculture, Plant Pest Control and
Laboratory Services Division, Library.** Capitol Building, Charleston,
WV 25305

Established: –
Stock: *c*. 2400 volumes, pamphlets, 55 current periodical titles
Main subjects: entomology, plant pathology, regulatory control
analysis, agriculture, chemistry

USSR

**1098 Nauchno-Issledovatel'skij Institut Prirodnoochagovych Infekcij,
Nauchnaja Medicinskaja Biblioteka** [Institute of Infectious Diseases].
Prospekt Mira 7, Omsk 80

Established: 1947
Stock: *c*. 15 000 volumes, 160 current periodical titles
Main subjects: infectious diseases, including relevant aspects of medi-
cal and veterinary entomology

**1099 Uzbekskij Nauchno-Issledovatel'skij Institut Eksperimental'noj
Meditsinskoj Parazitologii i Gel'mintologii im. L.M. Isaeva, Nauchnaja
Biblioteka** [Uzbek Institute for Research in Experimental Medical
Parasitology and Helminthology]. Isaeva 38, Samarkand 5

Established: 1924
Stock: *c*. 55 000 volumes, 210 current periodical titles
Main subjects: medical parasitology and helminthology, including
relevant aspects of medical entomology

1100 Vsesoyuznoe Entomologicheskoe Obshchestvo Bibliotek [All-Union

Entomological Society Library]. Universitetskaya Naberezhnaya, Leningrad V-164

Established: 1860
Stock: *c*. 30 000 items, 15 000 current periodical titles
Main subjects: entomology

DATES OF PUBLICATION

The correct dating of natural history publications, particularly those carrying descriptions of new taxa, is of great importance. Accurately established dates of publication are essential in settling issues of priority in nomenclature. A considerable amount of research has been done in the past to establish the dates of issues of parts of these works, and of those published as a whole. Many of these published papers are buried in journals and are difficult to find. However, references have been systematically indexed in a series of papers in the *Journal of the Society for the Bibliography of Natural History*. That Society continued to collect such references, as do many of the major natural history libraries; for example the libraries of the British Museum (Natural History).

1101 **Griffin, F.J., Sherborn, C.D.** and **Marshall, H.S.** 1936. A catalogue of the papers concerning the dates of publication of natural history books. *Journal of the Society for the Bibliography of Natural History* 1: 1–30.

1102 **Griffin, F.J.** 1943. Catalogue of papers concerning the dates of publication of natural history books. Supplement 1. *Journal of the Society for the Bibliography of Natural History* 2: 1–18.

1103 **Stearn, W.T.** and **Townsend, A.C.** 1953. Catalogue of papers concerning the dates of publication of natural history books. Supplement 2. *Journal of the Society for the Bibliography of Natural History* 3: 5–12.

1104 **Goodwin, G.H.** 1957. Catalogue of the papers concerning the dates of publication of natural history books. Supplement 3. *Journal of the Society for the Bibliography of Natural History* 3: 165–74.

1105 **Goodwin, G.H., Stearn, W.T.** and **Townsend, A.C.** 1962. Catalogue of papers concerning the dates of publication of natural history books. Supplement 4. *Journal of the Society for the Bibliography of Natural History* 4: 1–19.

These works are indexed by author, date and abbreviated title, with a full bibliographic reference to the paper in which the dating of the particular book is discussed.

6 Keeping up with current events

NEWSLETTERS

Newsletters are much neglected as a source of information in literature searching. They are now being produced in ever-increasing numbers in many areas of research, and the subject of entomology is no exception. They are produced for very specialized areas of the subject, and usually on a voluntary basis. Most of these newsletters would not be regarded, either by the compilers or the contributors, as 'available publications'. They are often sent out free or on a cost-recovery basis only.

The newsletter, however, should be regarded as a most important source of information. Newsletters usually carry current bibliographies, typically giving the readers access to references well in advance of any of the abstracting and indexing services. Research projects are often described and in many cases, newsletters are also a rich source of biographical information. Many carry names and addresses of research workers or lists of subscribers.

All titles in the following list were, as far as we have been able to ascertain, current at the time of going to press. We have tried to make the list as complete as possible – but many are difficult to obtain. Most of the titles are self-explanatory of the groups that they cover.

1106 **Amateur Entomologists' Society. Exotic Entomology Group Newsletter.** 4 per year. Source: B. Morris, 34 Borden Lane, Sittingbourne, Kent ME1 1DB, England.

Short notes from contributors; 'wants and exchanges' of specimens. Some book reviews.

1107 **Amateur Entomologists' Society. Insect Behaviour and Ants Study Group Newsletter.** Source: M. Parson, 95 Glebe Lane, Barming, Maidstone, Kent, England.

Short articles and observations, reports of meetings; names and addresses of members.

1108 **Aphidologist's Newsletter.** Biannual. Source: Dr C.S. Wood-Baker, 10 Green Lane, Chislehurst, Kent, England.

Short articles, exchanges of specimens; book reviews. Current literature and lists of aphid workers with their addresses.

1109   Balfour-Browne Club. Source: Dr G.N. Foster, 20 Angus Avenue, Prestwick, Ayrshire KA9 2HZ, Scotland.

News and information of meetings. Distribution maps. Short research and collecting notes. Brief comments on recently published papers and books.

1110   Biological Weed Control Newsletter. Irregular. Source: D. Schroeder, CIBC Swiss Station, Delémont, Switzerland.

Research notes, news, reports and recent publications on the biological control of weeds.

1111   Buprestis. Irregular. Source: Hans Mühle, Schlossermauer 45, D-8900 Augsberg, Federal Republic of Germany.

Short notes and reviews. Exchanges of information; names and addresses of members.

1112   Ceratopogonidae Information Exchange. 2 per year. Source: J. Boorman, Animal Virus Research Institute, Pirbright, Woking, Surrey, England.

Current research news; bibliographies of current and retrospective literature. Available to all interested workers.

1113   Chironomus. Source: E.J. Fittkau, Zoologische Staatssammlung, Maria-Ward-Strasse 16, D-8000 München 19, Federal Republic of Germany.

Short notes and research communications. Current bibliographies. Biographical material and occasional book reviews.

1114   Chrysomela. Source: T.N. Seeno, Department of Food and Agriculture, 1220 N. Street, Room 340, Sacramento, California 95814, USA.

Research articles, lists of workers on the subject with addresses. Bibliographies of current literature.

1115   Coccidologists' Newsletter. Source: D.R. Millar, Systematic Entomology Laboratory, 11B111, AR, USDA, Building 003, Rm 1, Beltsville, Maryland 20705, USA.

Short research articles. Current literature and names and addresses of workers on Coccids.

1116   Curculio. Source: H.R. Burke, Department of Entomology, Texas A & M University, College Station, Texas 77843, USA.

Current research news and literature reviews. Names and addresses of subscribers.

1117   Drosophila Information Service. Source: Division of Biological Sciences, University of Kansas, Lawrence, Kansas 66045, USA.

Extensive current literature lists. Research notes and lists of available stocks of specimens. Lists of workers with addresses.

1118 Eatonia. Source: Florida A & M University, Box 78 A, Tallahassee, Florida 32907, USA.

The declared aim of the newsletter is 'to acquaint all workers with current research of others; promote increased knowledge of the literature, especially among workers recently entering the field, and to promote more precise methods and techniques of studying Ephemeroptera'. Occasional supplements are issued containing larger papers on Ephemeroptera.

1119 Entomological Livestock Group. Source: J.D. Stewart, 73 Guisborough Road, Nunthorpe, Cleveland TS7 0JS, England.

Fortnightly lists of livestock, mainly of unusual/exotic species of Lepidoptera and other arthropods.

Entomology Newsletter – *see* entry 445

1120 Flea News. Source: R.E. and J.H. Lewis, Department of Entomology, Iowa State University, Ames, Iowa 50011, USA.

Occasional bibliographies. Current literature, research news and addresses of research workers.

1121 Frass Newsletter. 2 per year. Source: Ronald Wheeler, Chevron Chemical Corporation, 940 Hensley Street, Richmond, California 94804, USA.

Published by the Insect Rearing Group; includes research notes, lists of suppliers for rearing equipment and media. Current bibliographies and lists of members.

1122 Heteropterists' Newsletter. Source: C.W. Schaeffer, Biological Science Group, University of Connecticut, Storrs, Connecticut 06268, USA.

Lists of works on Heteroptera, special bibliographies and current literature, information on collections and research programmes.

1123 Ichnews (Ichneumonid newsletter). Source: M.G. Fitton, Department of Entomology, British Museum (Natural History), Cromwell Road, London SW7 5BD, England.

Current research news and literature.

1124 International Cecidology Newsletter. 2 per year. Source: Entomology Research Unit, Loyola College, Madras, India.

Carries news about research activities, and notes on current literature. Obituaries and biographical notes are often included.

1125 IOBC Newsletter. Source: Secretary-General IOBC, 1 rue de Nôtre, F-75016 Paris, France.

The newsletter of the International Organization for Biological Control of Noxious Animals and Plants. Includes news, reports, abstracts of relevant papers published in IOBC's journal *Entomophaga*.

1126   IPM Practitioner. 12 per year. Source: Bio Integral Resource Center, Rt 1, Box 28A Wintress, California 95694, USA.

Deals with all aspects of integrated pest management. Includes research notes, book and product reviews, lists of suppliers (for biological control organisms, etc.), letters and calendar of events.

1127   Locust Newsletter. Source: Plant Production and Protection Division, Food and Agriculture Organization of the United Nations, Rome, Italy.

Short articles, reports of commissioners and agencies. News of travel of headquarters staff.

1128   Newsletter of the British Simulium Group. Source: T.R. Williams, Department of Zoology, The University, PO Box 147, Liverpool L69 3BR, England.

'To maintain and develop contact between those interested in Simuliidae.' Exchange of news; exchange of information concerning all aspects of Simuliid biology. Current research, lists of names and addresses of workers on the subject.

1129   Newsletter of the Balfour–Brown Club. Source: M. Sinclair, Lynnwood Lodge, 10 Liddesdale Road, Hawick, Roxburghshire TD9 0ES, Scotland.

Short notes and current research on Coleoptera. Names and addresses of interested subscribers.

1130   News of the Lepidopterists' Society. Source: Dr D. Winter, 257 Common Street, Dedham, Massachusetts 02026, USA.

A well-established newsletter. Issues a membership list every two years. Research and collecting notes; some book reviews.

1131   Perla. A Newsletter for Plecopterists. Source: R.W. Baumann, Department of Zoology, Brigham Young University, Provo, Utah 84602, USA.

Current research notes and literature. Lists of subscribers and research workers on Plecoptera.

1132   Pest Management News. Source: UC/AID Pest Management Project, 137 Ciannini Hall, University of California, Berkeley, California 94720, USA.

News of developments in all aspects of pest management. Worldwide calendar of events and meetings.

1133   Phasmid Study Group Newsletter. Source: T. James, 'Manzil', Trewithen Terrace, Greatwork, Ashton, Helston, Cornwall TR13 9TQ, England.

Exchange of information between members and exchanges of livestock.

1134   Proctos. Proctotrupoid Newsletter. Source: L. Masner, Biosystematics

Research Institute, Agriculture Canada, Research Branch, Ottawa, Ontario, Canada.

Recent publications, news of members and research notes.

1135 Scarabaeus. Source: A.R. Hardy, c/o Insect Taxonomy Laboratory, California Department of Food and Agriculture, 1220 N. Street, Sacramento, California 95814, USA.

Short notes and bibliography of recent literature.

1136 Selysia. A Newsletter of Odonatology. Source: M.J. Westfall, Department of Zoology, University of Florida, Gainesville, Florida 32611, USA.

Current research and research notes. Current literature lists.

1137 Sphecos. A Newsletter for Aculeate Wasp Researchers. Source: A.S. Mencke, Systematic Entomology Laboratory, USDA, ARS, c/o US National Museum of Natural History, Washington, DC 20560, USA.

Research notes, current literature. Extensive biographical notes on members.

1138 Sting. Source: J. Woets, Glasshouse Crops Research and Experimental Station, Postbus 8, NL-2670 Naaldwijk AA, Netherlands.

Concerned with all aspects of biological control in glasshouses. Includes regional reports, meeting reports, news items and recent publications.

1139 Trichoptera Newsletter. Source: Hans Malicky, Biologische Station, A-3293 Linz, Austria.

Research notes, current literature and the names and addresses of research workers and subscribers.

CONFERENCE CALENDARS

1140 Antenna. Bulletin of the Royal Entomological Society. 4 per year. Royal Entomological Society of London, 41 Queen's Gate, London SW7 5HU, England.

Lists forthcoming meetings and conferences, not only of the Society, but of other organizations. Will list forthcoming conferences, symposia, etc. well into the future, provided a firm date has been set and a venue.

1141 Calendar of UK Biological Societies. 12 per year. Institute of Biology, 41 Queen's Gate, London SW7 5HU, England.

Divided into two main sections. Part 1 lists British biological societies, with addresses and membership numbers. The second part is the calendar, which lists month by month meetings that have been notified to the Biological Council.

1142 Forthcoming International Scientific and Technical Conferences. 4 per year. Aslib, 3 Belgrave Square, London SW1X 8PL, England.

Each part has an index to organizations sponsoring a listed conference. There is also an index by location and a subject index. The information provided is the date, title of conference, the location and the contact point.

1143   Index of Conference Proceedings Received. 12 per year. British Library, Lending Division, Boston Spa, Wetherby, West Yorkshire LS23 7BQ, England.

Lists conference proceedings, available for loan, received by the British Library Lending Division. The work is arranged by subject keywords. Within the subject the references are arranged chronologically by date of meeting.

1144   International Conferences Related to Pest Management. 12 per year. Compiled by the Pest Management and Related Environmental Protection Project at Berkeley, California, USA, under USAID contract AID/DSAN-C-0252 with the Consortium for International Crop Protection (CICP).

Listings can be found for conferences projected well into future years, sometimes as much as four to five years ahead.

7 Entomologists and their organizations

LEARNED SOCIETIES AND MEMBERSHIP LISTS

Many countries of the world can offer to those interested an Entomological Society. Some are large and internationally famous; others rather smaller and perhaps less active. Many of them publish their own journal, or commercial publishers edit and sell one on their behalf.

We have listed only those that are devoted entirely to entomology, but it should be remembered that Natural History Societies frequently have entomological sections, or certainly entomologists among their membership.

There is no World Directory of entomologists as there is for some professions. However, many societies do issue lists of members at irregular intervals. These can be useful sources of addresses and details of members' interests.

Every four years, entomologists of the world meet at the International Congress of Entomology, and the list of participants that is invariably produced as part of the Congress literature is a most useful source for addresses. The list is often accompanied by a booklet describing the position of entomology in the host country. This usually lists members of the profession, societies, institutions and university departments employing entomologists.

Directories of entomologists

AFRICA

1145 Ojal, J.M. 1979. *Directory of insect scientists of Africa.* International Centre of Insect Physiology and Ecology; Nairobi, Kenya.

An alphabetical list by name and interest, institutes being indicated by abbreviations, followed by a list of workers under their countries of origin, and finally an index of abbreviations with the full title and address.

1146 Kumar, R. (Compiler). 1976. *Directory of Ghana's entomologists.* Zoology Department, University of Ghana; Legon.

The directory is arranged alphabetically by name; it gives the address, field of specialization and current research interests.

1147 Youdeowei, A. (Ed.). 1975. *Directory of Nigerian entomology.* 32pp. Nigerian Entomological Society; Ibadan.

Included is an alphabetical list of members with their full names and addresses plus their field of research or interest.

AUSTRALIA

1148   Directory of Australian entomology. 1972. *Miscellaneous Publications of the Australian Entomological Society* No. **2**: 1–51.

An alphabetical list of Australian entomologists, including the full name and address, plus the area of interest. There is subject breakdown of fields of interest. The list includes state government departments and their staff; universities and colleges; and museums and non-government organizations concerned with entomology.

BELGIUM

1149 *Société Royale d'Entomologie – Liste des membres 1980.* 1980. 12pp. Société Royale d'Entomologie; Brussels.

An alphabetical list of names with full addresses, sent to all members of the Society. In some cases subjects of interest are included.

CANADA

1150 *Annotated list of workers on systematics and faunistics of Canadian insects and certain related groups.* 1977. 107pp. Entomological Society of Canada; Ottawa.

An alphabetical list indicating the name, address, position and major field of interest. Several indices are included: an index to locations of larger entomological establishments; index to taxa – individuals listed with current research projects; and an index to ecological groups.

EUROPE

1151 *Societas Europaea Lepidopterologica – News of the Society* No. **5**: 1–54.

A full list of members, with full addresses. A separate list includes names of members willing to do identifications. *News of the Society* No. **7** includes addenda and corrigenda to the membership list.

GHANA – *see* Africa

INDIA

1152 Jotwani, M.G. 1967. *Who's who in the Entomological Society of India, with directory of pesticides industry in the country.* 146pp. Entomological Society of India, New Delhi.

The volume includes the names and addresses of members of the Entomological Society of India; the entries include the employer and interests of the member.

INTERNATIONAL

1153 Abdullah, M. 1976. The higher classification of the insect order Coleoptera including fossil records and a classified directory of the coleopterists and Coleoptera collections of the world. *Zoologische Beiträge* **21**: 363–461.

The list includes curators and collectors known to offer collections on loan and important writers on phylogenetic systematics.

1154 Blackwelder, R.E. and **Blackwelder, R.M.** 1961. *Directory of zoological taxonomists.* 404pp. Southern Illinois University Press; Carbondale, Illinois.

Many entomologists are included, with full names and addresses; their special area of interest is added. Now badly outdated.

1155 Cummins, K.W. 1973. *A worldwide directory of stream ecologists.* 67pp. Institute of Water Research, Michigan State University, East Lansing, Michigan.

A number of specialists on aquatic entomology are included (e.g. Simuliidae), but they are few in number; the list is by no means complete.

1156 *Directory of scientists engaged in research on pathology of insects and other invertebrates.* 1967. Laboratory of Invertebrate Pathology, Department of Zoology and Entomology, Ohio State University and Ohio Agricultural Research and Development Centre; Columbus and Wooster, Ohio.

A numbered alphabetical name list with addresses and speciality subject. There is a subject index of principal research areas, and a list according to country. The numbers in the index refer back to the main directory.

1157 *International Congresses of Entomology.* No. 16. 1980. Kyoto, Japan. List of participants. 60pp.

The list includes some 2000 names and addresses of participants in the last International Congress of Entomology. No other information is given, but this is a most useful list of addresses.

1158 Wilson, M.R. (Compiler). 1981. *Directory of research workers on the Homoptera–Auchenorrhyncha.* 31pp. University College, Cardiff.

Research workers are listed alphabetically; they are also listed by interest and by country.

JAPAN

1159 *Entomology in Japan.* 133pp. XVIth International Congress of Entomology. 1980.

A most useful source for names and addresses of Japanese entomologists. The volume also includes a list of institutions and university departments concerned with entomology.

NEW ZEALAND

1160   Ramsay, G.W. and **Singh, P.** 1982. *Guide to New Zealand entomology.* 72pp. Entomological Society of New Zealand; Auckland. (*Bulletin of the Entomological Society of New Zealand* No. 7.)

Intended as a resource book for teachers, students, professionals and amateurs and deals with: history, achievements, resources, research, services, institutions and societies, education and careers, conservation and collecting, and sources of entomological materials. There is an annotated list of New Zealand literature, and a section on children's books. A directory of the Society's members is incorporated, and those who are willing to help with identification are indicated. A very useful idea, which other national societies might like to consider copying.

NIGERIA — *see* Africa

PACIFIC (including islands, etc.)

1161   Koehler, I.M. (Compiler). 1966. *Pacific entomologists.* Pacific Scientific Information Centre, Bernice P. Bishop Museum; Honolulu, Hawaii.

A list of entomologists actively interested in insects of the Pacific area. Names are listed alphabetically with addresses. An 'interest index' is included, arranged under subject, and a geographical list. The addresses are now somewhat outdated.

UK

1162   Amateur Entomologists' Society. Membership list.

Issued annually to members. The list includes names and addresses; it is also cross-referenced under regions. Interests of members are included.

1163   Association of British Zoologists. List of members. 1963. 21pp. Pergamon Press; London.

Names and addresses followed by the branch of zoology in which the member is interested.

1164   British Entomological and Natural History Society. Membership list.

Issued at irregular intervals. Includes full names and addresses.

1165   Royal Entomological Society of London. List of Fellows. 1980. 97pp.

An alphabetical list of Fellows with full addresses. The work includes an index of overseas members listed by country.

USA

1166   Entomological Society of America. List of members. 1980. *Bulletin of the Entomological Society of America* **26**: 190–318.

Lists of members with full names and addresses are published in the *Bulletin* at irregular intervals.

1167   Florida Entomological Society. 1981. *Handbook of Florida entomologists.* 53pp.

Lists of names and addresses of members, the posts they hold and areas of specialization of study. The list also includes institutions with entomological activities in Florida.

1168   Gupta, A.P. (Ed.) 1976. *Entomology in the USA.* v + 159pp. XVth International Congress of Entomology; Washington, DC.

A directory of entomologically orientated organizations. Departments of academic institutions, and private and commercial organizations are listed by state. The entomology staff with their speciality are listed for each organization.

1169   Hull, W.B. and **Odland, G.C.** (Eds). 1979. *Directory of North American entomologists and acarologists.* 188pp. Entomological Society of America; College Park, Maryland.

The directory is arranged by state; it also lists entomologists and acarologists working overseas, by country. It also includes an alphabetical list by speciality with reference back to the main part of the directory for address.

1170   Jacques, R.L. 1974. *Directory of American coleopterists, including descriptions of their research projects and a list of abbreviations of collection names for the world.* iii + 86pp. Biological Research Institute of America; Latham, New York.

Index of American coleopterists with full names and addresses (usually those of the institutions to which they are attached). Research projects are listed.

Entomological societies

ARGENTINA

1171   Sociedad Entomológica
         Argentina
         Calle Maipú 267
         Buenos Aires

AUSTRALIA

1172   Australian Entomological
         Society
         c/o Department of Primary
         Industries
         Meiers Road
         Indooroopilly
         Queensland 4068

1173 Entomological Society of
Australia (NSW)
PO Box 22
Five Dock, New South Wales
2046

1174 Entomological Society of
Queensland
c/o Entomology Department
University of Queensland
St Lucia
Brisbane, Queensland 4067

1175 Entomological Society of
Victoria
c/o Zoology Department
La Trobe University
Bundoora, Victoria 3083

AUSTRIA

1176 Arbeitsgemeinschaft für
Ökologische Entomologie
in Graz
Ludwig Boltzmann-
Institut für Umweltwissen-
schaften und Naturschutz
Heinrichstrasse 5/III
A-8010 Graz

1177 Arbeitsgemeinschaft Öster-
reichischer Entomologen
Ludo Hartmann-Platz 7
A-1160 Wien

BELGIUM

1178 Société Royale
d'Entomologie de
Belgique
rue Vautier 31
B-1040 Bruxelles

1179 Union des Entomologistes
Belges
M.L. Berger
34 avenue du Manoir
d'Anjou
Bruxelles 15

1180 Vereniging voor
Entomologie van de
Koninklijke Maatschappij
voor Dierkunde van
Antwerpen
Diksmuidelaan 176
B-2600 Berchem

BRAZIL

1181 Sociedade Brasileira de
Entomologia
Caixa Postal 9063
01000 São Paulo, SP

CANADA

1182 Entomological Society of
Canada
1320 Carling Avenue
Ottawa K1Z 7K9

1183 Entomological Society of
British Columbia
B.D. Frazer
6660 NW Marine Drive
Vancouver, British
Columbia V6T 1X2

1184 Entomological Society of
Ontario
D.H. Pengelly
Environmental Biology
University of Guelph
Guelph, Ontario

1185 Entomological Society of
Quebec
M.C. Bouchard
Complexe Scientifique,
D-1-59
2700 rue Einstein, Ste-Foy
Québec 12

1186 Entomological Society of
Saskatchewan
Agriculture Canada
Research Station
Saskatoon, Saskatchewan

CHILE

1187 Sociedad Chilena de
Entomología
Casilla 4150
Santiago

COLOMBIA

1188 Sociedad Colombiana de
Entomología
Apartado Aéreo 24718
Bogotá DE

CZECHOSLOVAKIA

1189 Československá Společnosti
Entomologická
Viničná 7
128 44 Praha

1190 Slovenská Entomologická
Společnosti
ul. Obrancov mieru 41
Bratislava

DENMARK

1191 Entomologisk Forening i
København
c/o Zoologisk Museum
Universitetsparken 15
DK-2100 København

1192 Lepidopterologisk Forening
Skorskellet 35A
DK-2840 Holte

EGYPT

1193 Société Entomologique
d'Egypte
PO Box 430
Cairo

FINLAND

1194 Lepidopterologiska
Sällskapeti Finland
P. Rautatiekatu 13
SF-00100 Helsinki 10

1195 Societas Entomologica
Helsingforsiensis
P. Rautatiekatu 13
SF-00100 Helsinki 10

FRANCE

1196 Société Entomologique de
France
45 rue de Buffon
F-75005 Paris

1197 Société Entomologique de
Mulhouse
35 place de la Réunion
F-68100 Mulhouse

1198 Société Entomologique du
Nord de la France
50 avenue des Lilas
F-59000 Lille

FEDERAL REPUBLIC OF GERMANY

1199 Arbeitsgemeinschaft
Rheinisch-Westfälischer
Lepidopterologen
Löbbecke-Museum und
Aquarium
Postfach 1120
D-4000 Düsseldorf

1200 Deutsche Gesellschaft für
Allgemeine und
Angewandte Entomologie
c/o Institut für
Phytopathologie
Ludwigstrasse 23
D-6300 Giessen

1201   Entomologischer Verein
       Apollo
       K.G. Schurian
       Altkönigstrasse 14a
       D-6231 Sulzbach/Ts

1202   Münchner Entomologische
       Gesellschaft (EV)
       Maria-Ward-Strasse 1b
       D-8000 München 19

1203   Internationaler
       Entomologischer Verein,
       Frankfurt-am-Main
       Senckenberg Museum
       Senckenberganlage 25
       D-6000 Frankfurt-am-Main

1204   Societas Europaea Lepidop-
       terologica
       Landessammlungen für
       Naturkunde Karlsruhe
       Postfach 4045
       D-7500 Karlsruhe 1

GREECE

1205   Hellenic Entomological
       Society
       PO Box 1925
       Thession
       Athens

HUNGARY

1206   Magyar Rovartani Társaság
       Baross u. 13
       H-1088 Budapest

INDIA

1207   Entomological Society of
       India
       Division of Entomology
       Indian Agricultural
         Research Institute
       New Delhi 12

1208   Association for the Study of
       Oriental Insects
       c/o Department of Zoology
       University of Delhi
       Delhi 7

INDONESIA

1209   Entomological Society of
       Indonesia
       Central Research Institute
       Agriculture Institute
       Bogor

IRAN

1210   Entomological Society of
       Iran
       PO Box 3178
       Tehran

ISRAEL

1211   Entomological Society of
       Israel
       c/o Volcani Institute of
       Agricultural Research
       PO Box 6, Bet-Dagan

ITALY

1212   Associazione Romana di
       Entomologia
       c/o Museo Civico di
       Zoologia
       Via Allise Aldrovandi, 18
       I-00197 Roma

1213   Accademia Nazionale
       Italiana di Entomologia
       via Romana 15–17
       I-50125 Firenze

1214   Società Entomologica Italiana
       via Brigata Liguria 9
       Genova

JAPAN

1215 Entomological Society of
Japan
c/o Department of Zoology
National Science Museum
(Natural History)
3-23-1 Hyakunin-chô
Shinjuku-ku
Tokyo 160

1216 Japanese Coleopterological
Society
c/o No. 199, 1–3 Nishitakaai
Higashisumiyoshi
Osaka 546

1217 Japanese Heterocerists'
Society
c/o National Science
Museum
Ueno Park
Tokyo

1218 Japanese Hymenopterists'
Society
Asahigaaka 4–15
Mishima 411

1219 Japanese Society for
Applied Entomology and
Zoology
c/o Japan Plant Protection
Association
Komagome
Toshima-ka
Tokyo 170

1220 Japanese Society for Sani-
tary Zoology
c/o The Institute of Medical
Science
University of Tokyo
Shiroganedai
Minatoku, Tokyo

1221 Lepidopterological Society
of Japan
c/o Entomological Labora-
tory
National Science Museum
(Natural History)
3-23-1 Hyakunin-chô
Shinjuku-ku
Tokyo 160

1222 Shikoku Entomological
Society
Entomological Laboratory
College of Agriculture
Ehime University
Matsuyama

1223 Society of Odonatology
c/o Takadanobaba
4-4-24 Shinjuku-ku
Tokyo 160

KOREA

1224 Entomological Society of
Korea
Korean Entomological Insti-
tute
Korea University
1 Anam-Dong, Sungbuk-ku
Seoul 132

MEXICO

1225 Sociedad Mexicana de
Entomología
Apartado Postal 7-1080
Delegación Cuauhtémoc
06700 México, DF

1226 Sociedad Mexicana de
Lepidopterología
Nicolas San Juan 1707
México 20, DF

NETHERLANDS

1227   International Odonatological Society
c/o Department of Animal Cytogenetics and Cytotaxonomy
University of Utrecht
Padualaan 8
Utrecht

1228   Nederlandse Entomologische Vereniging
c/o B.J. Lempke
Plantage Middenlaan 64
NL-1004 Amsterdam

NEW ZEALAND

1229   Entomological Society of New Zealand
Entomological Division
DSIR, Private Bag
Auckland

NORWAY

1230   Norsk Entomologisk Forening
PO Box 7508
Skillebek, Oslo 2

1231   Norsk Lepidopterologisk Selskap
Zoologisk Museum
Sarsgt. 1
Oslo 5

PERU

1232   Sociedad Entomológica de Peru
Apartado 4796
Lima

PHILIPPINES

1233   Philippine Association of Entomologists
c/o Department of Entomology
University of the Philippines
College
Laguna 3720

POLAND

1234   Polskie Towarzystwo Entomologiczne
ul. Sienkiewicza 21
50-335 Wrocław

SOUTH AFRICA

1235   Entomological Society of Southern Africa
PO Box 103
Pretoria

SPAIN

1236   Asociación Española de Entomología
c/o Departamento de Zoología
Facultad de Ciencias
Universidad de Salamanca
Patio de Escuelas 1
Apartado 20
Salamanca

1237   Instituto Español de Entomología
J. Gutiérrez Abascal 2
Madrid 6

1238   Sociedad Astúrica de Lepidopterología
Calle Bérmudez de Castro 19
Coriedo

1239 Sociedad Hispano-Luso-
Americana de Lepidop-
terología
Torre de Madrid
Suite 5-12
Madrid 13

1240 Societat Catalana de
Lepidopterologia
Apartat de Correus 13
Mataró (Maresme)
Catalunya

SWEDEN

1241 Entomologiska Föreningen i
Stockholm
Zoologiska Institutionen
Helgonavägen 3
S-223 62 Lund

SWITZERLAND

1242 Entomologische
Gesellschaft Basel
Faberstrasse 1
Postfach 70
CH-4047 Basel

1243 Entomologische
Gesellschaft Luzern
Natur-Museum Luzern
Kasernenplatz 6
CH-6003 Luzern

1244 Schweizerische
Entomologische
Gesellschaft
c/o W. Meier
Eidg. Forschungsanstalt
Beckenholz
CH-8046 Zürich

UK

1245 Amateur Entomologists'
Society
108 Hanover Avenue
Feltham
Middlesex TW13 4VP

1246 British Entomological and
Natural History Society
c/o The Alpine Club
74 South Audley Street
London W1

1247 Cheshire and Lancashire
Entomological Society
68 Bridgenorth Road
Pensby
Wirral, Merseyside

1248 Royal Entomological
Society of London
41 Queen's Gate
London SW7 5HU

USA

1249 American Entomological
Society
Academy of Natural
Sciences
1900 Race Street
Philadelphia, Pennsylvania
19103

1250 Association of Applied
Insect Ecologists
c/o PO Box 712
Exeter, California 93221

1251 Cambridge Entomological
Club
16 Divinity Avenue
Cambridge, Massachusetts
02138

1252 Connecticut Entomological
Society
c/o Department of Biology
Yale University
PO Box 6666
New Haven, Connecticut
06520

1253 Entomological Society of
America
4603 Calvert Road
College Park, Maryland
20740

1254   Entomological Society of
       Pennsylvania
       Division of Forest Pest
       Management
       Pennsylvania Department of
       Environmental Resources
       Middletown, Pennsylvania
       17057

1255   Entomological Society of
       Washington
       c/o Department of
       Entomology
       Smithsonian Institution
       Washington, DC 20560

1256   Florida Entomological
       Society
       Florida Department of
       Agriculture and Consumer
       Services
       Division of Plant Industry
       PO Box 1269
       Gainesville, Florida 32602

1257   Georgia Entomological
       Society
       USDA – SEA – AR
       Southern Grain Insects
       Research Laboratory
       Tifton, Georgia 31794

1258   Hawaiian Entomological
       Society
       c/o Department of
       Entomology
       University of Hawaii
       3050 Maile Way
       Rm 310
       Honolulu, Hawaii 96822

1259   Kansas Entomological
       Society
       c/o Department of
       Entomology
       University of Kansas
       Lawrence, Kansas 66045

1260   Lepidopterists' Society
       1041 New Hampshire Street
       Lawrence, Kansas 66044

1261   Lepidoptera Research
       Foundation
       1160 W. Orange Grove
       Avenue
       Arcadia, California 01006

1262   Maryland Entomological
       Society
       1215 Stella Drive
       Baltimore, Maryland 21207

1263   Michigan Entomological
       Society
       c/o Department of
       Entomology
       Michigan State University
       East Lansing, Michigan
       48824

1264   Mississippi Entomological
       Society
       Extension Entomology
       PO Box 5426
       Mississippi State 39762

1265   New York Entomological
       Society
       American Museum of
       Natural History
       Central Park West at 79th
       Street
       New York 10024

1266   North Carolina Entomologi-
       cal Society
       North Carolina State Uni-
       versity
       Raleigh 27607

1267   Pacific Coast Entomological
       Society
       California Academy of
       Sciences
       Golden Gate Park
       San Francisco
       California 94118

1268 South Carolina Entomological Society
Department of Entomology and Economic Zoology
Clemson University
Clemson 29631

1269 Southwestern Entomological Society
Department of Entomology
Texas A & M University
College Station 78843

1270 Tennessee Entomological Society
Department of Entomology and Plant Pathology
University of Tennessee
Knoxville, Tennessee 37901

1271 Washington State Entomological Society
Department of Entomology
Washington State University
Pullman, Washington 99164

1272 Wisconsin Entomological Society
Department of Entomology
University of Wisconsin
Madison, Wisconsin 53706

1273 Xerxes Society
Briggs Hill Road
Sherman, Connecticut 06784

URUGUAY

1274 Sociedad Uruguaya de Entomología
Casilla de Correo 490
Montevideo

USSR

1275 Vsesoyuznoe Entomologicheskoe Obshchestvo
Zdadnie Zoologicheskaya Instituta
Universitetskaya nab 1
19916 Leningrad B-164

1276 Latvijas Entomologijas Biedribas
Latvijas PSR
Zinātnu Akademijas
Botaniskais Darzs
Riga, Latvia

PROFESSIONAL ORGANIZATIONS

Although entomological interests are well catered for by the many societies that exist around the world, there is virtually no entomological equivalent of the type of certifying body found in most other disciplines. There are however a few organizations that go at least some way towards filling this gap:

1277 **American Registry of Professional Entomologists (ARPE).** 4603 Calvert Road, PO Box AJ, College Park, Maryland 20740, USA.

Founded: 1971
Membership: *c.* 1300
Publications: *News* (4 per year); *Registry Notes* (4 per year)
Aims and activities: ARPE is a certifying body aiming to help in the identification of qualified individuals who provide technical services to the public in entomologically related activities. Provides a continuing registry (certification has to be renewed annually) of qualified personnel engaged in pest management consultancy, pest con-

trol advisers, those providing special services including identification and the marketing of biotic agents, and those who are trained and competent in some related professional speciality. Assists educational institutions in the development of courses consistent with the standards required for certification.

1278   Association of Applied Biologists. Department of Agricultural Botany, University College of Wales, Aberystwyth, Dyfed SY23 3DD, Wales.

Founded: 1904
Membership: *c.* 1200 (UK); 200 (overseas)
Publications: *Annals of Applied Biology* (9 per year)
Aims and activities: Advancement of all branches of biology with special reference to their applied aspects. Ordinary and student memberships are open to all suitably qualified workers in applied biology, irrespective of nationality. Activities include indoor and field meetings.

1279   Institute of Biology. 20 Queensbury Place, London SW7 2DZ, England.

Founded: 1950
Membership: *c.* 15 000
Publications: *Biologist* (5 per year); *Journal of Biological Education* (4 per year); *Symposium Reports* (1 per year); *Careers in biology* (booklet); *Studies in biology* (monographic series); *Organizations in the UK Employing Biologists*; *Directory of Independent Consultants in Biology*; *Calendar of Meetings of Biological Societies*
Aims and activities: Sets professional standards for biology, and generally represents the profession nationally and internationally. There are six grades of membership: Fellow (FIBiol), Member (MIBiol), Graduate, Associate, Affiliate and Student. Admission to the first four of these is generally dependent on academic qualification, or by successful completion of the Institute's own graduate examination.

1280   Royal Entomological Society of London. 41 Queen's Gate, London SW7 5HU, England.

Founded: 1833
Membership: *c.* 1500 Fellows (FRES)
Publications: *Ecological Entomology* (4 per year); *Physiological Entomology* (4 per year); *Systematic Entomology* (4 per year); *Symposia* (1 per 2 years); *Handbooks for the identification of British insects* (monographic series); *Antenna* (4 per year); *List of Fellows*
Aims and activities: Holds regular meetings and biennial international symposia. Maintains an extensive and historically important library. The great majority of professional entomologists in the UK are Fellows, as are many overseas.

OBITUARIES AND BIOGRAPHIES

Obituary notices and biographies can be rich sources of information. Not only will they tell the inquirer some of the more obvious information, such as the dates of birth and death (although sometimes, it is this most obvious requirement that is left out), but can also be a useful source for the inquirer on collections. They may say where such collections were deposited, and perhaps mention the areas where the deceased travelled and collected specimens. They may also tell of changing interests or areas of research. These published records are often accompanied by bibliographies and sometimes portraits.

There are in existence many sources for biographical information. Most countries publish national biographical dictionaries, but few individuals reach the professional heights that make for inclusion in publications such as these. There are several sources specifically on entomologists, and it is these that are mostly given here. A few more general but useful sources are also included. Histories of the subject should not be ignored; these often include biographical material, especially for the more famous names. Another rich source is the history of a learned society or of a geographical area. Used together these can often provide a mass of useful information.

It is recommended that this section be used in conjunction with that on the histories of the subject; notes are appended to those references which include biographical information.

1281 *Biographical memoirs of the Fellows of the Royal Society.* 1955–. Preceded by *Obituary notices of the Fellows of the Royal Society.* 1932–54.

These provide very full biographies of deceased Fellows of the Royal Society. Each is accompanied by a photograph and a full bibliography of publications. The volumes are issued annually.

1282 *Biographical memoirs of the National Academy of Sciences.* 1877–.

Extensive biographies and bibliographies, usually accompanied by a photograph, published on the death of members of the Academy.

1283 Carpenter, M. 1945, 1953. Bibliography of biographies of entomologists. *American Midland Naturalist* **33**: 1–116; **50**: 257–348.

Deceased entomologists are listed alphabetically; full details of dates are given where known and references to obituaries and biographical material.

1283A Conci, C. 1975. Repertorio delle biografie e bibliografie degli scrittori e cultori italiani di entomologia. *Memorie della Società Entomologica Italiana* **48** (1969): 817–1078.

Entomologists are listed alphabetically with full bibliographic references to obituaries and biographies.

1284 Darby, M. 1980–. *A biographical dictionary of British coleopterists.* To be issued in parts. Pages 1–4 issued with *Coleopterist's Newsletter* No. 6 (1981).

1285   Gilbert, P. 1977. *A compendium of the biographical literature on deceased entomologists.* 455pp. British Museum (Natural History); London.

An alphabetical list of deceased entomologists (or persons who occupied themselves with entomology at some time during their lives) accompanied by their full names and dates where known. There is a chronological list of references to known published biographies or obituaries. References to published bibliographies and portraits are also given.

1286   Judd, W.W. 1979. *Early naturalists and natural history societies of London, Ontario.* 215pp. Phelps; London, Ontario.

Much biographical material is included in this volume; many of the individuals included were entomologists.

1287   Mallis, A. 1971. *American entomologists.* xvii + 549pp. Rutgers University Press; New Brunswick, New Jersey.

This is a good source book for biographical material. A very high percentage of the better-known names in American entomology are included in this volume, as well as many portraits.

1288   Musgrave, A. 1932. *Bibliography of Australian entomology 1775–1930, with biographical notes on authors and collectors.* 380pp. Royal Zoological Society of New South Wales; Mosman, NSW.

A good source for early Australian zoologists and entomologists, both research workers and collectors. Brief biographies are given, along with biographical references. Short bibliographies are also included, where they refer to Australian material.

1289   *Obituaries from 'The Times'. Including an index to all obituaries and tributes appearing in 'The Times'.* 3 volumes. Newspaper Archive Developments Ltd; London.
1951–1960   Published 1979
1961–1970   Published 1975
1971–1975   Published 1978

An alphabetical sequence of reprinted obituaries appearing in *The Times*, London. Each volume has an index of names giving the reference to the day of publication, page and column number in the newspaper. This is an invaluable source of information, particularly for the well-known names and also those working in associated areas such as medicine, etc.

1290   *Obituary notices of the Fellows of the Royal Society.* 1932–54.

Annual volumes carrying extensive biographies of deceased Fellows of the Royal Society. Full bibliographies of published papers accompany the notice. The series continues as *Biographical Memoirs of the Fellows of the Royal Society 1955–* (entry 1281).

1291   Papavero, N. 1971–6. *Essays on the history of Neotropical dipterol-*

ogy, with special reference to collectors (1750–1905). 3 volumes. Universidade de São Paulo. Museu de Zoologia; São Paulo.

The title of this work is somewhat misleading, since it leads the reader to presuppose that the people discussed in the volumes worked only with Diptera. This is not so, and the volume is a most useful source for biographical information on many of the early collectors and naturalists in South America.

1292 Rohlfien, K. 1977. Bibliographie entomologischer Bibliographien (1920–1970). *Beiträge zur Entomologie* **27**: 313–79.

Part of this most useful work has a good section on personal bibliographies.

8 Miscellaneous services

It is perhaps inevitable in a work of this kind that a few items should find themselves quite unjustly relegated to the last few pages, simply because they do not fit into any convenient category. This is of course no reflection on their importance, and entomologists may console themselves with the knowledge that one of the most widely used library classification schemes actually places all literature on librarianship in a catch-all 'generalities' class!

TRANSLATIONS: SERVICES AND GUIDES

Scientists are not always good linguists. It is one thing to acquire a small but select vocabulary enabling one to pick out a few relevant words from a research paper, but to have a good understanding of a piece of research done and written in a language other than one's own is a different matter, and can cause problems. Contracting for translations can be an expensive business. Many scientists have friends who will oblige, but others may find it necessary to budget for funds to have translations prepared on demand. To prevent the duplication of this expensive work, many countries have set up deposit holdings for translations, and information is supplied to a central organization for indexing. Such centres include the National Translations Centre in the USA, the British Library in the UK and the International Translations Centre in the Netherlands. These and similar organizations deal mainly with translations of articles from scientific journals; however, there also exist programmes that have undertaken the translation of certain works on a cover-to-cover basis. The Smithsonian Institution and the Israel Program for Scientific Translation have undertaken complete volumes of *Fauna of the USSR* and sold them on a commercial basis. Up to 1981 such works were also translated by the British Library Lending Division in the UK.

Some organizations such as Aslib (Association of Special Libraries and Information Bureaux) keep a register of translators, and although they do not undertake translations themselves, can put a subscriber in touch with a suitable translator.

1293 **Aslib (Association of Special Libraries and Information Bureaux).** 3 Belgrave Square, London SW1X 8PL. *Directory of translators and translation services.* Published annually.

Aslib maintains the Commonwealth Index of Unpublished Scientific and Technical Translations. No translations are held, but information is available of organizations that may hold copies of the translation in question. Copies of the index are held and maintained by organizations in Commonwealth countries, i.e. CSIRO, Australia; National Research Council, Canada, and the Indian National Science Documentation Centre, India. No translations are undertaken by Aslib, but its staff can put subscribers in touch with translators – sometimes a subject specialist.

1294 Bees. International Bee Research Association, Hill House, Chalfont St Peter, Gerrards Cross, Buckinghamshire SL9 0NR, England. *Apicultural Abstracts. Bee World.*

Both publications are issued quarterly. *Apicultural Abstracts* lists available translations on bees. *Bee World* includes translations added to the Association's library.

1295 British Library Lending Division. Boston Spa, Wetherby, West Yorkshire LS23 7BQ, England. *British Reports, Translations and Theses Received by the British Library Lending Division (Including Material from the Republic of Ireland).* Monthly list.

British report literature and translations produced by British government organizations, industry, universities and learned institutions.

Most guides list the British Library as willing to undertake translations, both of whole volumes and of articles in scientific and technical journals. This service ceased in 1981. The Library does, however, collect translations and make them available for loan.

1296 British Library Science Reference Division. 25 Southampton Buildings, London WC2A 1AW.

This division of the British Library does not produce or publish translations. However, it will provide, by appointment, a brief oral translation for library users. This may give the user a brief indication of the content of a paper, so that he or she may decide whether it is necessary to contract for a complete translation.

1297 Commonwealth Agricultural Bureaux. Farnham House, Farnham Royal, Slough SL2 3BN, England.

The CAB have recently set up a translation service run by CAB staff who, as well as having considerable expertise in their particular area of study, also have linguistic qualifications. Prices are determined for each piece of work done, currently falling within the range of £20–£35 per thousand words of original text. At present European languages are covered including some Scandinavian and Slavonic languages and the service will probably extend to Arabic, Chinese and Japanese.

1298 Centre National de la Recherche. 88 rue St Dominique, Paris 8e, France.

Translates some 4500 articles *per annum*, mostly on demand. Translations are undertaken from all languages into French and from French to other languages. Photocopying and microfilm services are available.

1299 ELB Translations Services Ltd. 61 Carey Street, London WC2A 2JG, England. Suite 907, 1119 Jefferson Davis Highway, Arlington, Virginia 22202, USA.

A commercial translation service with offices in the UK and USA. 'A worldwide professional service for all foreign-language needs.' The company claims to translate any language and subject and will try to use subject specialist and language combination where possible.

1300 Index Translationum. *International bibliography of translations.* Unesco. Issued annually. Vol. 30 (1981).

The current volume is a bibliographical catalogue of some 50 500 translated books and published in 1977 in the member states. The work is arranged by countries in alphabetical order and presented under 10 major headings of the Universal Decimal Classification, the natural and exact sciences being listed under heading 5.

1301 International Translations Centre. 101 Doelenstraat, Delft, The Netherlands. *World Transindex.* Published monthly.

Announces translations of literature relating to all fields of science and technology from Eastern European and Asiatic languages. Translations from Western European languages into French, Spanish and Portuguese are also announced. Copies are available for sale, and a photocopy service is provided. The Centre does not undertake translations, but maintains a register of translators both by specialist language and subject specialization. Most national translation centres have committed arrangements for notifying the International Translations Centre of deposits to national centres.

1302 National Translations Center. The John Crerar Library, 35 West 33rd Street, Chicago, Illinois 60616, USA.

The principal US depository for unpublished translations into English of the natural, physical, medical and social sciences. The estimated growth is some 10 000 *per annum*. The Center undertakes no translations work, but maintains a directory of translators. Translations are available for loan or by photocopy. The Center has a reciprocal arrangement for loan and photocopy service with the British Library Lending Division, UK (entry 1295).

World Transdindex – *see* **International Translations Centre** (entry 1301)

APICULTURAL INFORMATION

Although to a non-specialist it may seem illogical, apiculture is generally regarded as a subject more or less separate from mainstream entomology. Although we have included elsewhere in this work a number of references to its literature, we have really only touched the tip of this particular iceberg; there is an enormous wealth of specialized apicultural literature and it is with some regret that we have been unable to include more.

However, those wishing to delve deeper into this fascinating subject are well provided for by the excellent publications and other services of the:

1303   International Bee Research Association. Hill House, Gerrards Cross, Buckinghamshire SL6 0NR, England.

BIOLOGICAL CONTROL

Although not strictly speaking within the remit of this volume, biological control continues, quite rightly, to arouse considerable interest among those involved in applied entomology. Its importance is evident from the tremendous proliferation of new books and other literature on the subject over the last few years. We have listed a small selection of these in other sections of this work, but readers who wish to pursue the subject further may need to seek specialist advice.

A number of organizations within the government, academic and commercial sectors are actively engaged in biological control work. One that is particularly well qualified to give advice, or to plan and implement actual control programmes, is the Commonwealth Institute of Biological Control. Originally founded in 1927 as the Farnham House Laboratory, CIBC is now part of the Commonwealth Agricultural Bureaux and has stations in many parts of the world. Correspondence should be addressed to either of the following:

1304   Commonwealth Institute of Biological Control. Gordon Street, Curepe, Trinidad, West Indies (for the attention of the Director).

1305   Commonwealth Institute of Biological Control. Imperial College Field Station, Silwood Park, Ascot, Berkshire SL5 7PY, England (for the attention of the Assistant Director).

Index

References are to item numbers, not pages, except as otherwise indicated.

computerized information retrieval,
pp. 155–7
Conci, C., 336, 1283A
conference calendars, 1126, 1132,
1140–4
conference proceedings, 1143
*Contributions of the American
Entomological Institute*, 415
Coope, R. G., 708
Coppell, H. C., 904
Corbet, P. S., 128, 609
Cornwell, P. B., 889–90
Costa Lima, A., 90
Coton et Fibres Tropicales, 416
cotton pests, 416, 867
Counce, S. J., 679
Coutin, R., 341
Crawford, C. S., 610
Crawford, L. D., 160
crickets, *see* Orthoptera
crop loss assessment, sugarcane, 883
Crosby, T. K., 254; p. 37
Crosskey, R. W., 71
Crowe, T. J., 170
Crowson, R. A., 48, 611, 709
Crozier, R. H., 664
CSIRO, 612
cultures, laboratory, 235–44
Cummins, K. W., 1155
Cummins, R. W., 117, 684
curation (*see also* collections), 245–54
Curculionidae, newsletter, 1116
Current Contents, 929
Current Research, 417
Cushing, E. C., 4
cuticle techniques, 765
cytogenetics (*see also* genetics), 664–8
Czechoslovakia
aquatic entomology, 629
entomological societies, 1189–90
fauna, 353, 355
libraries, 1003–5

D'Abrera, B., 84
Dahms, E., 246
Dalton, S. C., 697
Daly, H. V., 613
Danks, H. V., 139–40
Darby, M., 1284
Darlington, A., 831
Darlington, C. D., 666
Darsie, R. F., 114
Data Sheets on Quarantine Organisms,
418

databases, *see* online information
retrieval
dates of publication, 1101–5
Davies, H., 785
Davis, D. D., 51
Davis, E. D., p. 9
DeBach, P., 906
defence (*see also* venoms), 658
Delfinado, M. D., 74
Della Beffa, G., 187
DeLong, D. M., 608
Denmark
entomological societies, 1191–2
general entomology, 439
Lepidoptera, 509
lepidopterists' society, 1192
Denno, R. F., 647
Derbena-Ukhova, V. P., 786
Derksen, W., 27
Dermaptera, *see* Orthoptera
desert invertebrates, biology and
ecology, 610
Dethier, V. G., 760
Deutsche Entomologische Zeitschrift,
419
Devecht, J. van, 851
developmental stages and embryology,
678–85
Deyrolle, 277
DIALOG information retrieval system,
955
Dickerson, W. A., 235
dictionaries (*see also* glossaries),
205–34
agriculture, 223
aquatic entomology (Russian), 224
ecology, 210
French, 227
genetics, 218
German, 214, 217
Russian, 228
diets
artificial, 237
natural, 758
Dingle, H., 647
Diptera
Afrotropical, 71
behaviour, 657
biology, 630
black-flies, 122
bloodsucking, 657
British, 127
disease vectors, 793
evolution, 723